MICROBIOLOGY RESEARCH ADVANCES

MODELS IN BIOSCIENCE AND MATERIALS RESEARCH

MOLECULAR DYNAMICS AND RELATED TECHNIQUES

MICROBIOLOGY RESEARCH ADVANCES

Additional books in this series can be found on Nova's website under the Series tab.

Additional e-books in this series can be found on Nova's website under the e-book tab.

MATERIALS SCIENCE AND TECHNOLOGIES

Additional books in this series can be found on Nova's website under the Series tab.

Additional e-books in this series can be found on Nova's website under the e-book tab.

MICROBIOLOGY RESEARCH ADVANCES

MODELS IN BIOSCIENCE AND MATERIALS RESEARCH

MOLECULAR DYNAMICS AND RELATED TECHNIQUES

KHOLMIRZO T. KHOLMURODOV
EDITOR

New York

For permission to use material from this book please contact us:
Telephone 631-231-7269; Fax 631-231-8175
Web Site: http://www.novapublishers.com

Additional color graphics may be available in the e-book version of this book.

LIBRARY OF CONGRESS CATALOGING-IN-PUBLICATION DATA

ISBN 978-1-62808-052-0
Library of Congress Control Number: 2013940405

Published by Nova Science Publishers, Inc. † New York

CONTENTS

PREFACE

The latest trends and developments in bioscience and materials research show that computer modeling combined with atomistic and molecular simulation approaches is gaining ground. Computer molecular dynamics modeling and experimental research are mutually beneficial, and their combination is fruitful practically in all branches of today's technonological innovations, medicine, and material design. Computer molecular simulation methods broaden the range of tools for the evaluation of material and drug properties and establishment of new (not discovered yet experimentally) ones before mass production. Specially mentioned should be biological chemistry, chemical engineering, and rapidly growing molecular biosciences and bioengineering, which promise to gain very much from molecular dynamics modeling concerning the treatment of, or understanding, the molecular mechanisms of different diseases faced by people today.

Computer molecular simulation is a powerful technique which allows one to investigate various physical or biological processes at the atomic level. X-ray or neutron measurements of physical, chemical, and biological structures inevitably involve the further use of computer models based on molecular and atomic simulation techniques. With the invention of modern supercomputers, the methods of computer molecular simulation (conventional or hybrid molecular dynamics (MD), Monte-Carlo, *ab initio* quantum chemistry, etc.) of large molecular systems have quickly shown their power. Molecular simulation (conventional and hybrid MD) is based on classical Newtonian physics, modeling the particle interaction in molecules via the force fields defined in advance – empirically or calculated by other methods. The novel computational approaches allow treating important problems in material fabrication and biomedical applications. Powerful modern computer-based molecular simulation methods became traditional tools in the industy of new materials and drugs. In this book, original papers are collected that demonstrate efficient uses of MD simulation for studying nanoscale phenomena in a number of models from material and life sciences. This volume contains the Proceedings of the 5th Japan–Russia International Workshop "Molecular Simulation Studies in Materials and Biological Sciences" (MSSMBS'2012), which was held at the Joint Institute for Nuclear Research (JINR), Dubna, and at the Institute of Bioorganic Chemistry (IBC) of the Russian Academy of Sciences, on September 09–12, 2012. MSSMBS'12 started at the JINR International Conference Hall, Dubna, and continued and finished at the IBC in Moscow.

The series of MSSMBS meetings started in 2004 and continued in 2006–2012. MSSMBS'04 was the first international conference held in Russia to be focused on methodological problems and applications of the art of MD simulations in physical, chemical, and biological systems. The idea of the MSSMBS workshop was first proposed by Prof. Kenji Yasuoka from Keio University, Japan, who is a famous expert in molecular dynamics simulations for chemical physics and biology. The MSSMBS workshops are mostly contributed by leading research groups of Japan and Russia and are also participated by scientists of USA, Canada, Europe, India, Mongolia, Egypt, and other countries. The subjects of the MSSMBS workshops include different aspects of molecular simulation in materials science and biological research; computational and theoretical studies of atomic and molecular interactions; dynamics between atoms, molecules, ions, clusters, and surfaces; modern high-performance computing facilities; and simulation techniques and methods applied for studying molecular systems and structures.

The MSSMBS'12 scope included, in particular, the following topics:

- Molecular dynamics simulation of proteins
- Protein folding research by generalized ensemble techniques
- Mutation transition effects in protein structures
- Molecular dynamics and Monte-Carlo simulations of radiation-induced mutations

- Modeling genetic regulatory networks in bacterial and mammalian cells
- Simulation of DNA damage induction by ionizing radiations
- Chemical and nanostructure design (crystal, liquid, and polymer systems)
- Drug design (docking of one molecule to another, enzymes, and inhibitory activities)
- Novel MD computing methods (DFT, QM/MM, MD, MD/CFD, and hybrid approaches)
- General: & special-purpose MD machines and modern communication architectures

The MSSMBS International Workshops provide an excellent opportunity for young and leading scientists of Japan, Russia, European and many other countries to meet and share their thoughts on the latest trends in the art. The MSSMBS workshops also aim to promote a new field of science at JINR, Dubna, which has traditionally been a nuclear research center performing nuclear physics and radiobiological experiments.

Kholmirzo T. Kholmurodov
Chairman
MSSMBS Organizing Committee

Joint Institute for Nuclear Research
6 Joliot-Curie St., Dubna, 141980, Moscow Region, Russia
E-mail: mirzo@jinr.ru

Molecular Dynamics Models in Biosciences and Materials Research: An Introduction

On September 9–12, the Laboratory of Radiation Biology of the Joint Institute for Nuclear Research (JINR, Dubna, Moscow Region, Russia) and the Institute of Bioorganic Chemistry of Russian Academy of Sciences (Moscow, Russia) hosted the 5th Japan–Russia International Workshop MSSMBS-2012 "Molecular Simulation Studies in Material and Biological Sciences" (co-chaired by Profs. Kholmirzo Kholmurodov and Roman Efremov). The MSSMBS'12 provided a lot of interesting talks on different aspects of materials science and biological research and outlined the current state and prospects of computer molecular simulation studies; it covered a broad spectrum of problems in modern physics, biochemistry, and nanotechnology. In particular, the following topics were discussed: molecular dynamics simulations of proteins; protein-folding research by generalized-ensemble techniques; mutation transition effects in protein structures; molecular dynamics and Monte-Carlo simulations of radiation-induced mutations; modeling genetic regulatory networks in bacterial and mammalian cells; simulation of DNA damage induction under ionizing radiations; chemical and nanostructure design (crystal, liquid and polymer systems); drug design (docking of one molecule to another, enzymes, inhibitory activities); novel MD computing methods (DFT, QM/MM, MD, MD/CFD, hybrid approaches); and general- & special-purpose MD machines and modern communication architectures. The MSSMBS'12 participants represented leading computer molecular simulation research groups of Japan (RIKEN, Okazaki Institute for Molecular Science, Waseda University, Nagoya University, Keio University, etc.), Russia (JINR, M.V. Lomonosov Moscow State University, St.Petersburg State University, Institute of Bioorganic Chemistry, the Joint Institute for High Temperatures of the Russian Academy of Sciences), Canada (University of Saskatchewan), France (University Paris-Diderot), etc.

In the first paper by S. Ohmura, K. Nagaya, F. Shimojo and M. Yao *ab initio* molecular-dynamics study was performed for nonequilibrium phenomena in disordered materials. The energy transport in light harvesting dendrimer and the coulomb explosion of bromophenol are studied by *ab initio* molecular-dynamics simulations. The thermal molecular motion plays an important role in electron transport in the light harvesting dendrimer. The simulation reveals that tetrahydrofuran solvent suppresses the molecular motion and consequently the electron transfer. The dissociation mechanism of a charged bromophenol is also investigated. It is clarified that the hydrogen atom of hydroxyl group dissociates from the molecule before the aromatic ring is broken.

In the the second paper by E. Dushanov, Kh. Kholmurodov, and K. Yasuoka the photoreaction and adsorption properties on surfaces, thermal decomposition, chemical transformation, and other properties of the formamide molecule are widely used to understand the origins of the formation of biological molecules (nucleosides, amino acids, DNA, monolayers, etc.) needed for life. The titanium dioxide (TiO_2) surface can act both as a template on which the accumulation of adsorbed molecules like formamide occurs through the concentration effect, and as a catalytic material that lowers the activation energy needed for the formation of intermediate products. In this paper, a formamide–water solution interacting with TiO_2 (anatase) surface is simulated using the molecular dynamics method. We also studied the structure and diffusion properties of the Ethanol/Water mixture, which play important roles in the theoretical study and technological applications involving mass transfer. The structural, diffusion and density properties of mixture on surface are established for a wide temperature range from $T = 250$ K up to $T = 400$ K.

In the 3rd paper by T. Yoda, Y. Sugita, and Y. Okamoto the results of a simulation study on folding of villin headpiece subdomain HP36 were reported. The multicanonical replica-exchange molecular dynamics simulations of HP36 in the presence and the absence of KCl salt ions in the explicitly represented water solvent were performed. During the production runs, each of which is shorter than the experimental folding time, the authors successfully observed folding events of HP36 to its native-like conformations. The salt effects on salt bridges within the protein molecule and discuss about their influence on the hydrophobic-core formation of the globular mini protein were examined.

In the 4th paper by O. Titov, D. Shulga, V. Palyulin, N. Zefirov anisotropic electrostatics approaches in description of halogen bonding within intermolecular potentials were studied. In the study the two most promising approaches to description of electrostatic part of XB - extra-point charge and multicenter multipolar distribution - are compared in different contexts: (1) in the reference molecular electrostatic potential (MEP) reproduction, (2) in the reference potential energy surface reproduction (complemented with van-der-waals parameters from GAFF and Tripos force fields), and (3) in estimating the interactions of a small molecule with a protein. Additionally numerical and other aspects of the approaches are compared for the prospects of further development. On the basis of this study multicenter multipolar distribution approach appears to be the most promising for empirical description of electrostatic part of XB.

In the 5th paper by A. Shahinyan, L. Arsenyan, A. Poghosyan the Study of phase diagram of the surfactant/water system by molecular dynamics simulation was reported. In their paper the authors performed 300 ns Molecular Dynamics (MD) to study the formation of spherical, cylindrical and flat micelles in the micellar and liquid-crystalline phases of amphiphile /water system. Some important parameters have been obtained which characterize the dynamic structure of micelles. It was found that diffusion rate of amphiphilic molecules in spherical, cylindrical and flat micelles are significantly different of each other because of the specific conditions of localization of molecules in these structural units.

In the 6th paper by H. Okumura a review of the multibaric-multithermal molecular dynamics algorithm is presented. In this algorithm, two-dimensional random walks not only in the potential-energy space but also in the volume space are realized. This algorithm is a powerful tool to discuss the temperature dependence and pressure dependence of biomolecules. Applications to an alanine dipeptide and chignolin in explicit water solvent are presented. The partial molar enthalpy change and partial molar volume change of unfolding can be calculated. Illustrating typical local-minimum free-energy conformations, folding and unfolding pathways can be revealed. We can also discuss the mechanism of temperature and pressure denaturation.

In the 7th paper by A. Kuznetsov, P. Volynsky, and R. Efremov the free energy profiles for each type of interaction can be obtained based on a given spatial structure of the transmembrane (TM) dimer using molecular dynamics (MD) with umbrella sampling and potential of mean force (PMF) calculation. Glycophorin A (GpA) is one of the well-studied membrane proteins. There is a lot of experimental and theoretical data obtained, but little is known about the role of environment in GpA TM dimer formation. So, the main goal of the present work is to estimate contributions of all components (protein, lipids, water) in association of TM helices for the wild-type GpA TM dimer and its mutant form. Also, a new algorithm of residue-based energy decomposition was proposed and tested.

In the 8th paper by Y. Novakovskaya hydrogen-bond networks as structure-determining conjugated systems were investigated. Analysis of the electron density distribution in clusters composed of hydrogen fluoride and water molecules reveals the existence of both σ- and π-binding between molecules. The σ-kind density distribution determines the mutual orientation of molecules. A π-system may be delocalized conjugated, which provides additional stabilization of molecular clusters. HF_2^- and $H_5O_2^+$ ions are characterized by quasi-triple bonds between the electronegative atoms. The most long-lived species stabilized by π-binding are structural rings and closed or open hoops and strips composed of fused rings. It is conjugated quasi-aromatic π systems that determine the cooperativity phenomenon. Spatial extension and structure of conjugated π systems determine the typical features of liquid hydrogen fluoride and water.

The 9th paper by T. Kobori, T. Otsuka, S. Tsuneyuki and M. Taiji presents a study on the development and application of fragment-based electronic structure calculation method for biomolecule. In this study the authors introduce an electronic structure calculation method for biomolecules, which we call FMO-LCMO method [1-2]. The term 'FMO' comes from Fragment Molecular Orbital [3-5]method, whose purpose is to calculate total energy and density matrix of the entire biomolecule with little computational cost. The reduction of CPU cost is accomplished by dividing the whole system into small subsystems called fragment. The FMO-LCMO method was developed by combining the FMO method with the idea of Linear Combination of Molecular Orbitals (LCMO). Just as Hamiltonian of any molecule is expressed with LCAO in conventional MO method, FMO-LCMO Hamiltonian is expressed with fragment molecular orbitals of each fragment. This property enables us to elucidate electronic structure of the whole system in terms of unit molecules such as amino acids or nucleotide. Then molecular orbitals and their orbital energies of the whole system can be calculated by diagonalization of the FMO-LCMO Hamiltonian. As our recent progress, we show how to calculate mulliken population with FMO-LCMO, and its calculation results for Chignolin with explicit water solvent. Charge distribution of both Chignolin and water molecules are also discussed.

In the 10th paper by R. Eremin, Kh. Kholmurodov, V. Petrenko, L. Rosta, M. Avdeev the molecular dynamics simulations were performed for small-angle neutron scattering: scattering length density spatial distributions for mono-carboxylic acids in d-decalin. The method of the neutron scattering length density (SLD) spatial distributions calculations based on the molecular dynamics simulation data is described on the example of saturated mono-carboxylic acids in decalin solutions. The sufficient SLD modulation in the solute-solvent interface area is obtained. Results of the current work and described calculation method can be used for small-angle neutron scattering data analysis.

In the 11th paper by J. Yamagishi, N. Okimoto, T. Kasai, A. Suenaga, M. Okada, A. Imamoto, and M. Taiji a computational design of peptide inhibitors, *in silico* method, has been discussed to design drug-like small molecules. In this study, the authors have developed docking-based protocols for the first-round screening. They developed our new docking program based on Molecular Mechanics and an implicit solvent model. The authors used a GPGPU technique to accelerate our calculations. They also considered the ligand's reorganization effect. This effect corresponds to the energetic contribution of the conformational change of the ligand through binding to the target protein. The authors used a small training set of phosphopeptides against CRK SH2 domain, which is a target for new cancer therapy. It was shown a high screening performance in discriminating between binding and non-binding sequences. The methods develped in this paper will also be helpful for more accurate design of small molecule inhibitors in the future.

In the 12th paper Kh. Kholmurodov, E. Dushanov, E. Krasavin, H. Hassan, H. ElHabashy, A. Galal, N. Sweilam, and K. Yasuoka have performed molecular dynamics (MD) simulations of the DNA interaction with metallic nanoparticles and TiO_2 surfaces. The understanding of the mechanism of DNA interactions and binding with metallic nanoparticles (NPs) and surfaces represents a great interest in today medicine applications due to diagnostic and treatment of oncology diseases. Recent experimental and simulation studies involve the DNA interaction with highly localized proton beams or metallic NPs (such as Ag, Au, etc.), aimed on targeted cancer therapy through the injection of metal micro- or nanoparticles into the tumor tissue with consequent local microwave or laser heating. The effects of mutational structure changes in DNA and protein structures could result in destroying of native chemical (hydrogen) bonds or, on the contrary, creating of new bonds that do not normally exist there. The cause of such changes might be the alteration of one or several nucleotides (in DNA) or the substitution of specific amino acid residues (in proteins), that can brought to the essential structural destabilization or unfolding. At the atomic or molecular level, the replacement of one nucleotide by another (in DNA double-helices) or replacement of one amino acid residue by another (in proteins) cause essential modifications of the molecular force fields of the environment that break locally important hydrogen bonds underlying the structural stability of the biological molecules. In this paper, the MD simulations were performed on four DNA models and the flexibilities of the purine and pyrimidine nucleotides during the interaction process with the metallic NPs and TiO_2 surface were clarified.

In: Models in Bioscience and Materials Research ISBN: 978-1-62808-052-0
Editor: Kh. T. Kholmurodov

Chapter 1

Ab Initio MOLECULAR-DYNAMICS STUDY OF NONEQUILIBRIUM PHENOMENA IN DISORDERED MATERIALS

Satoshi Ohmura[1,*]***, Kiyonobu Nagaya***[1]***, Fuyuki Shimojo***[2]
and Makoto Yao[1]
[1]Department of Physics, Faculty of Science
Kyoto University, Japan
[2]Department of Physics, Faculty of Science
Kumamoto University, Japan

Abstract

The energy transport in light harvesting dendrimer and the coulomb explosion of bromophenol are studied by *ab initio* molecular-dynamics simulations. The thermal molecular motion plays an important role in electron transport in the light harvesting dendrimer. The simulation reveals that tetrahydrofuran solvent suppresses the molecular motion and consequently the electron transfer. The dissociation mechanism of a charged bromophenol is also investigated. It is clarified that the hydrogen atom of hydroxyl group dissociates from the molecule before the aromatic ring is broken.

Keywords:*Ab initio*, Molecular Dynamics, Energy Transport, Bromophenol, Tetrahydrofuran Solvent

---

*E-mail address: ohmura@scphys.kyoto-u.ac.jp

1. Introduction

The non-equilibrium phenomena in disordered materials are attracting great attention. In particular, electron transport and coulomb explosion in molecules are of paramount importance in understanding interactions between materials and optical field. Energy transport in synthetic supermolecules such as light-harvesting dendrimers are great important for the solution of the global energy problem. [1] In the dendrimers, the electronic excitation energy due to photoexcitation of antennas located on the periphery of the molecules is rapidly transported to the photochemical reaction centers at the cores of the molecules, which in turn performs useful work such as photosynthesis and molecular actuation. [2] There are a number of experimental [3, 4, 5] and theoretical [6, 7, 8, 9, 10] researches about the rapid energy transport mechanisms in light-harvesting dendrimers. Such energy transfer is conventionally attributed to either dipole-dipole interactions (Förster mechanism) [11, 12] or the overlapping of donor and acceptor electronic wave functions (Dexter mechanism). Recently, we reveals that the key role of thermal molecular motion that significantly accelerates the energy transport based on the Dexter mechanism in the dendrimer. [13]

Another interesting phenomena due to interactions between materials and optical field is coulomb explosion. When a molecule is irradiated by hard x-ray, holes are proliferated through the Auger cascade and the molecule takes highly charge. The highly charged molecular ion are quite unstable due to the coulomb repulsive force and typically destroyed on a subpicosecond time scale. The fragment ions may convey information about molecular shape because they reflects geometric structure of the parent molecule just before dissociation. Recently, Nagaya and Sugishima *et al.* investigated the dissociation of bromide and fluoride aromatic molecules by position-sensitive time-of-flight measurements. [14, 15] They found that the charged fragment ions are emitted almost in a single plane. However details of the dissociation mechanism of these molecules are still unknown.

In this paper, we report on detailed investigations of microscopic mechanisms of energy transport in a light-harvesting dendrimer and the coulomb explosion of a bromide aromatic molecule by *ab initio* molecular-dynamics (MD) simulations, in which interatomic forces are calculated quantum mechanically in the framework of density-functional theory (DFT). The purposes of our study are to confirm that the energy transfer mechanism is Dexter mechanism in the light-harvesting dendrimer under the calculation taking into account many-body effects which are not introduced directly in earlier work [13] and to clarify the effect of tetrahydrofuran (THF) solvent on rapid energy transport. We also clarify the atomic dissociation mechanism in the coulomb explosion of the highly charged bromophenol.

2. Method of Calculation

The electronic states are calculated using the projector-augmented-wave (PAW) method [16, 17] within the framework of DFT. The generalized gradient approximation (GGA) [18] is used for the exchange-correlation energy. The plane-wave cutoff energies are 30 and 250 Ry for the electronic pseudo-wave functions and the pseudo-charge density, respectively. The Γ point is used for Brillouin zone sampling. The equations of motion for atoms are solved via an explicit reversible integrator. [19] For the simulation of the dendrimer, the system consists of a zinc-porphyrin core and a benzyl ether-type antenna. A time step of $\Delta t = 0.48$ fs is used. The nonadiabatic MD simulations are carried out by incorporate electronic transitions through the fewest-switches surface-hopping (FSSH) method [20] along with Casida's linear-response time dependent density functional theory (LR-TDDFT), [21] using the ground-state kohn sham orbitals as a basis set. Here, many-body effects are introduced through coupling matrix elements consisting of the random-phase-approximation and exchange correlation terms involving the nonlocal Hatree-Fock exchange potential. The nuclei are treated classically in the adiabatic representation, i.e., the atomic forces are calculated from the (excited) electronic eigenstates for the current nuclear positions. Switching probability from the current adiabatic state to another is computed from the density-matrix elements obtained by solving the TDKS equations [22], and nonadiabatic transitions between adiabatic states occur stochastically.

For the simulation of the coulomb explosion, the system is the 4-bromophenol in which a bromine atom and hydroxyl group are located on the opposite sides of the aromatic ring. Initially, we perform isothermal and isovolume (NVT) molecular dynamics simulation for the neutral molecule at 300K for 2.4 ps. After the MD simulation, we made the system positively charged +6. The cluster simulations for charged system were carried out using cluster boundary conditions which were imposed using the method of Martyna and Tuckerman. [23] The method rigorously corrects the error resulting from infinite replication of an aperiodic system using a screening function in reciprocal space. A time step of $\Delta t = 0.048$ fs is used before a hydrogen atom dissociates from the molecule and then $\Delta t = 0.24$ fs is used in the cluster calculation.

3. Light-Harvesting Dendrimer

3.1. Electron Trasport Mechanism

Figure 1 shows an example of the time evolution of the excitation energy and spatial distribution of an pseudo electron density. Initially, the electron density is distributed mainly in the left periphery (Fig. 1(a)). At about 6 fs, a transition occurs, accompanied by the transfer of the electron to the right periphery (Fig. 1(b)) through the overlapping state (Fig. 1(c)). At about 17 fs, another transition occurs, causing the electron density to reside mainly within the core (Fig. 1 3(d)) through the another type of overlapping states Fig. 1(e). The crossings or approaching excitation energy due to thermal motions of atoms produces these overlapping states. With larger fluctuation excitation energy, the crossings and approaching are more frequent, resulting in fast energy transfer. From this calculation in which we take into account many-body effects, we conclude that the energy transfer mechanism is Dexter type mechanism in the light-harvesting dendrimer.

3.2. Effects of the Solvent on Energy Transfer

In the experiment, the anhydrous tetrahydrofuran (THF) is used as a solvent. We explicitly introduce 14 THF molecules around the dendrimer as shown in Fig. 2 and treat them quantum-mechanically as well to study the effect of the solvent used. The weak dispersion (van der Waals) interactions between the dendrimer and THF molecules are treated semiempirically, in the adiabatic MD simulations. The time evolution of electronic eigenenergies at 300 K is shown in Fig. 2, where both the results with and without the THF solvent are plotted. It is obvious that the solvent suppresses the fluctuation of eigenenergies. From this calculation, we conclude that the existence of THF solvent results in a significant drop in the energy-transfer efficiency.

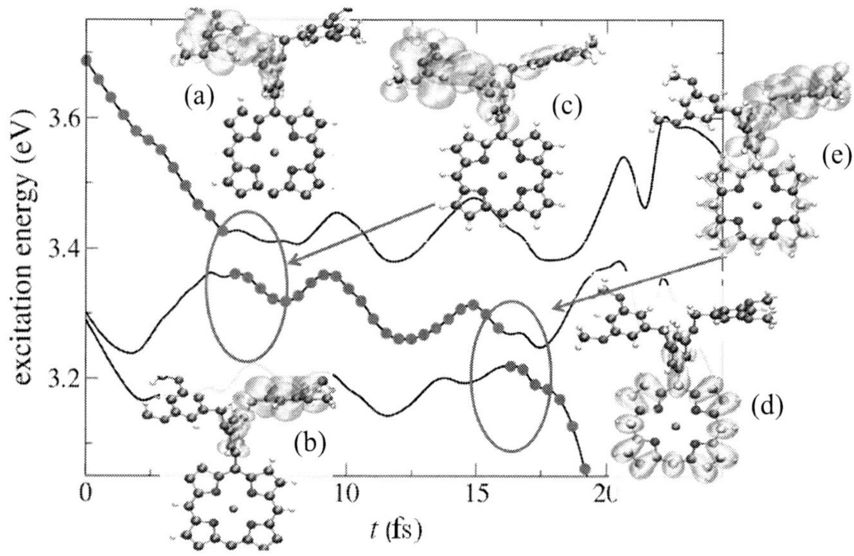

Figure 1. The time evolution of the excitation energy and the spatial distribution of the pseudo electron density.

4. Coulomb Explosion of Bromophenol

4.1. Hydrogen Dissociation

We used population analysis [24, 25] to clarify the dissociation mechanism of the charged bromophenol. By expanding the electronic wave functions in an atomic-orbital basis set, we obtained the overlap population O_{ij} between the ith and jth atoms and the gross charge Q_i for the ith atom. O_{ij} gives a semiquantitative estimate of the strength of the covalent-like bonding between atoms. Figure 3 shows the time evolution of the bond-overlap populations $O_{ij}(t)$. The value of $O_{O-H}(t)$ (the thick black line) decreases with time and becomes almost zero at about 20 fs. However the other $O_{ij}(t)$ have finite value at the same time, which means that the hydrogen atom of hydroxyl group dissociates from the molecules before aromatic ring is broken.

The time evolution of the gross charge Q_i is shown in Fig. 4. At $t = 0$ fs, the values of Q_{Br}, Q_H and Q_O are about 1.75, 0.75 and 0, respectively. The

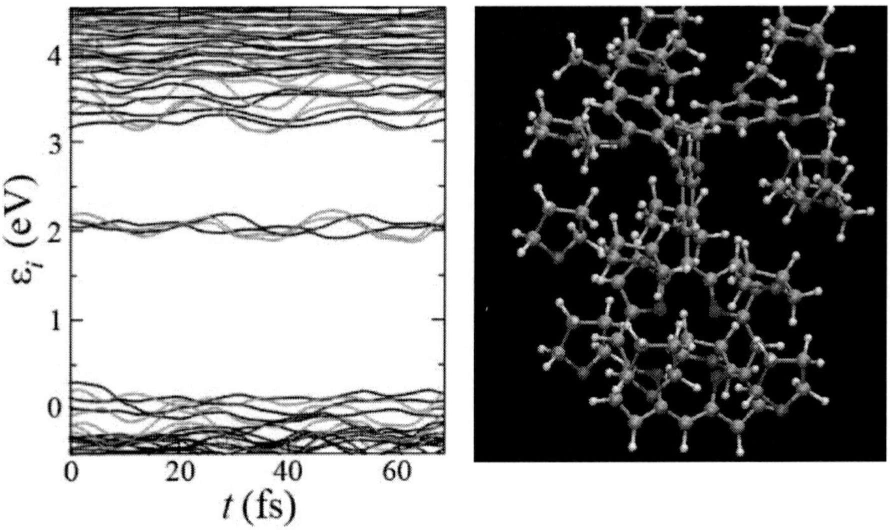

Figure 2. (Left panel) Time evolution of electronic eigenenergies during adiabatic MD simulation at 300 K. The black and red curves are obtained from MD simulations with and without THF solvent, respectively. (Right panel) A snapshot of atomic configuration. Dendrimer consisting of a zinc-porphyrin core with three benzenes and a benzyl ether-type antenna is surrounded by 14 THF molecules.

Q_H and Q_O increase with time, while the Q_{Br} decreases. At about 20 fs, Q_H is +1 means that hydrogen atom of hydroxyl group becomes proton after the dissociation.

4.2. Aromatic Ring-Opening

The electronegativity of the oxygen is largest in those of atomic species that make up the molecule. After the hydrogen atom of hydroxyl group dissociates from the molecules, oxygen intends to take electron from around next C atom including C-C bonds. As a result, the C-C bonds weaken and one of them is broken. Around Br atom, the situation is the same, which the electronegativity of Br atom is larger than that of C atom. Since the electron of C atom is taken by the Br atom, the one of C-C bonds next to Br atom is broken. This process is shown in Fig. 5.

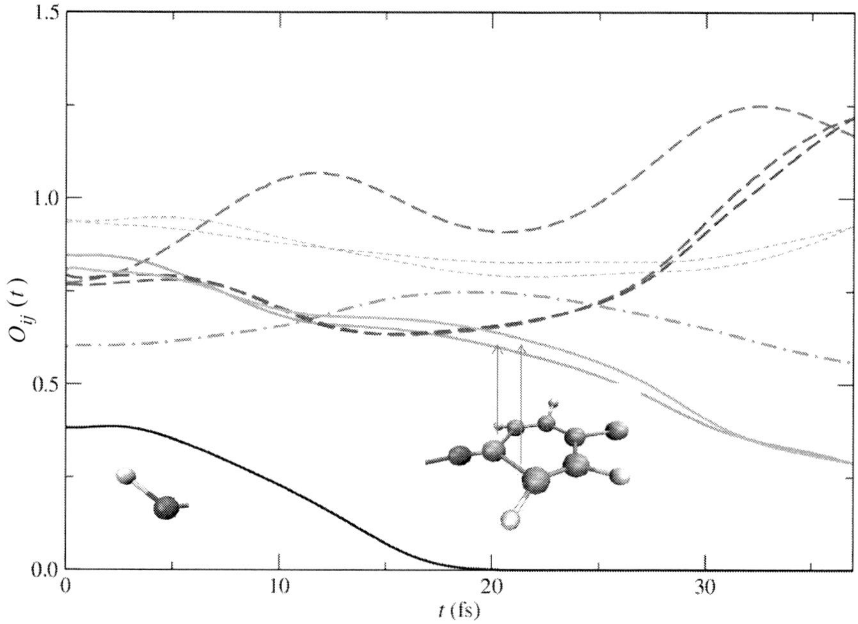

Figure 3. Time evolution of the bond-overlap populations $O_{ij}(t)$.

5. Summary

We have investigated the energy transport mechanism in light-harvesting dendrimer and the dissociation mechanism of the charged bromophenol by *ab initio* molecular-dynamics simulations. Our simulation clarified the energy transport mechanism in the dendrimer and the effects of the solvent used in experiments on the energy transfer efficiency. In the light-harvesting dendrimer, thermal molecular motion significantly accelerates the energy transport based on the Dexter mechanism. However, the anhydrous tetrahydrofuran (THF) solvent suppresses the molecular motion and consequently the electron transfer efficiency. For the +6 charged bromophenol, the dissociation mechanism is clarified. From the simulation, it is found that the hydrogen atom of hydroxyl group dissociates from molecules before the aromatic ring openning.

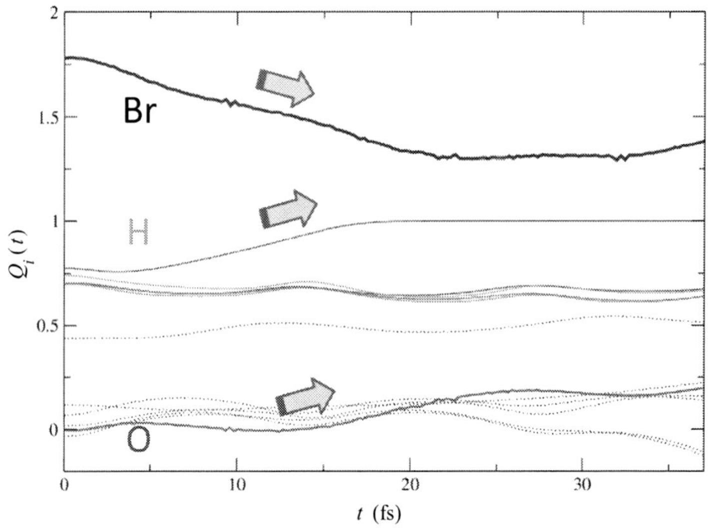

Figure 4. Time evolution of gross charge Q_i

Acknowledgment

The authors acknowledge useful discussions with Professor Kazuhiro Matsuda. This work was supported by KAKENHI [Grant-in-Aid for JSPS Fellows No. 24·5357]. The computations were performed using the computer facilities at the Supercomputer Center, Institute for Solid State Physics, University of Tokyo.

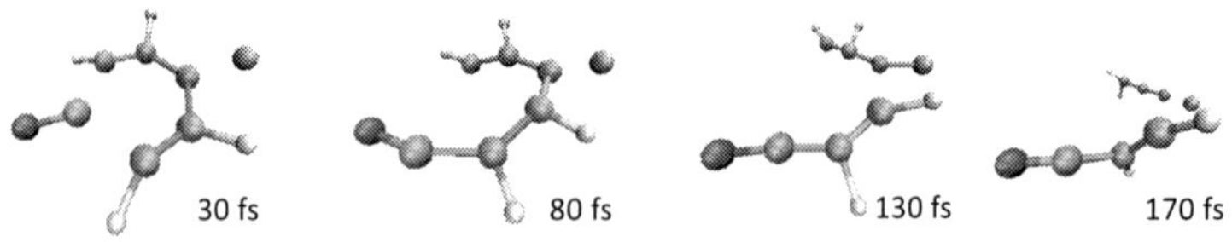

Figure 5. The ring opening process after hydrogen dissociation.

References

[1] N. S. Lewis, "Toward Cost-Effective Solar Energy Use". *Science* 315, (2007) 798.

[2] T. Muraoka, K. Kinbara, and T. Aida, "Mechanical twisting of a guest by a photoresponsive host". *Nature* 440, (2006) 512.

[3] I. Akai *et al.*, "Rapid energy transfer in light-harvesting small dendrimers". *J. Lumin.* 112, (2005), 449.

[4] A. Yamada *et al.*, "Energy Transfer Dynamics in Light-Harvesting Small Dendrimers Studied by Time-Frequency Two-Dimensional Imaging Spectroscopy". *J. Lumin.* 129, (2009), 1898.

[5] I. Akai *et al.*, "Rapid energy transfer in a dendrimer having p-conjugated light-harvesting antennas". *New J. Phys.* 10, (2008), 125024.

[6] R. Kopelman *et al.*, "Spectroscopic Evidence for Excitonic Localization in Fractal Antenna Supermolecules". *Phys. Rev. Lett.* 78, (1997), 1239.

[7] S. Raychaudhuri *et al.*, "Excitonic Funneling in Extended Dendrimers with Nonlinear and Random Potentials" *Phys. Rev. Lett.* 85, (2000), 282.

[8] K. Nishioka, and M. Suzuki, "Dynamics of unidirectional exciton migration to the molecular periphery in a photoexcited compact dendrimer". *J. Chem. Phys.* 122, (2005), 024708.

[9] R. Kishi *et al.*, "Core molecule dependence of energy migration in phenylacetylene nanostar dendrimers: *Ab initio* molecular orbital-configuration interaction based quantum master equation study". *J. Chem. Phys.* 128, (2008), 244306.

[10] Y. Kodama, S. Ishii, and K. Ohno, "Dynamics simulation of a -conjugated light-harvesting dendrimer II: phenylene-based dendrimer (phDG2)". *J. Phys.: Condens. Matter* 21, (2009), 064217.

[11] F. Vögtle, G. Richardt, and N. Werner, Dendrimer Chemistry (Wiley-VCH, Weinheim, 2009).

[12] A. Adronov, and J. M. J. Frechet, "Light-harvesting dendrimers". *Chem. Commun.* 18, (2000), 1701.

[13] S. Ohmura, S. Koga, I. Akai, F. Shimojo, R. K. Kalia, A. Nakano, and P. Vashishta "Atomistic mechanisms of rapid energy transport in light-harvesting molecules". *Appl. Phys. Lett.* 98, (2011), 113302.

[14] A. Sugishima, K. Nagaya, H. Iwayama, M. Yao, J. Adachi, Y. Kimura, M. Yamazaki, and A. Yagishita, "g Dissociation dynamics of C6H6 and C6H5F molecules following carbon 1s and fluorine 1s photoionization studied by three-dimensional momentum imaging method". *J. Chem. Phys.* 131, (2009), 114309.

[15] K. Nagaya, H. Iwayama, A. Sugishima, Y. Ohmasa, and M. YaO " Contactless measurements of charge migration within single molecules". *Appl. Phys. Lett.* 96, (2010), 233101.

[16] P. E. Blöchl, "Projector augmented-wave method". *Phys. Rev. B* 50, (1994), 17953.

[17] G. Kresse and D. Joubert, "From ultrasoft pseudopotentials to the projector augmented-wave method". *Phys. Rev. B* 59, (1999), 1758.

[18] J. P. Perdew, K. Burke and M. Ernzerhof, "Generalized Gradient Approximation Made Simple". Phys. Rev. Lett. 77, (1996), 3865.

[19] M. Tuckerman, B. J. Berne and G. J. Martyna, "Reversible multiple time scale molecular dynamics". *J. Chem. Phys.* 97, (1992), 1990.

[20] J. C. Tully, "Molecular dynamics with electronic transitions". *J. Chem. Phys.* 93, (1990), 1061.

[21] M. E. Casida, "in Recent advances in density functional methods (Part I),". edited by D. P. Chong (World Scientific, Singapore, 1995), p. 155.

[22] W. R. Duncan, C. F. Craig, and O. V. Prezhdo, "Time-domain *ab initio* study of charge relaxation and recombination in dye-sensitized TiO_2". *J. Am. Chem. Soc.* 129, (2007), 8528.

[23] G. J. Martyna, M. E. Tuckerman, "A reciprocal space based method for treating long range interactions in *ab initio* and force-?eld-based calculations in clusters". *J. Chem. Phys.* 110, (1999), 2810.

[24] R. S. Mulliken, "Electronic Population Analysis on LCAO-MO Molecular Wave Functions. I". *J. Chem. Phys.* 23, (1955), 1841.

[25] F. Shimojo, A. Nakano, R. K. Kalia, and P. Vashishta,"Electronic processes in fast thermite chemical reactions: A first-principles molecular dynamics study". *Phys. Rev. E* 77, (2008), 066103.

In: Models in Bioscience and Materials Research ISBN: 978-1-62808-052-0
Editor: Kh. T. Kholmurodov

Chapter 2

MD Studies of the Structural and Diffusion Properties of Formamide/Water and Ethanol/Water Mixtures on Titanium and Platinum Surfaces

Ermuhammad Dushanov[1,2,*]***, Kholmirzo Kholmurodov***[1,3]***, Kenji Yasuoka***[4]***, and Evgenii Krasavin***[1,3]
[1]Laboratory of Radiation Biology, JINR, Dubna, Moscow Region, Russia
[2]Institute of Nuclear Physics, Ulugbek, Tashkent, Uzbekistan
[3]Dubna International University, Dubna, Moscow Region, Russia
[4]Department of Mechanical Engineering, Keio University, Japan

Abstract

Formamide – surface interaction was one of the main discussion topics at the round table meeting "Astrobology: New Ideas and Research Trends" of the Italia – Russia@Dubna (2011). The nucleic acid components synthesized by formamide and their cosmic dust analogs have been widely considered to be the best candidates with regards to the early stages of the formation of biological molecules needed for life. Titanium dioxide (TiO_2, anatase), zircon (ZrO_2), and other minerals could

[*]E-mail address: dushanov@jinr.ru

influence as matrix elements the synthesis of nucleic acid bases from formamide, which is followed by temperature and UV heating. In the dark conditions, the experimental data on formamide adsorption at 300 K over the (001) plane of TiO_2 indicate some amount of unreacted formamide and water among the products (CO, H_2, NH_3, HCN). So far, some concentration of water and its involvement in formamide/TiO_2 interaction has been experimentally shown to be an important component of the process. In this paper, molecular dynamics (MD) simulation of interaction between a formamide – water solution and a TiO_2 (anatase) surface is performed. A comparative MD analysis is done of the formamide – water/(TiO_2, Pt) and ethanol – water/(TiO_2, Pt) systems in a temperature range from $T = 250$ K to $T = 400$ K to compare their structural and diffusion properties.

Keywords:Formamide Molecule, Ethanol Molecule, Water, Surface, Catalysts, Molecular Structures

1. Introduction

The photoreaction and adsorption properties on surfaces, thermal decomposition, chemical transformation, and other properties of the formamide molecule are widely used to understand the origins of the formation of biological molecules (nucleosides, amino acids, DNA, monolayers, etc.) needed for life. The titanium dioxide (TiO_2) surface can act both as a template on which the accumulation of adsorbed molecules like formamide occurs through the concentration effect, and as a catalytic material that lowers the activation energy needed for the formation of intermediate products. Formamide contains the four elements (C, H, O, and N) most required for life and it is attractive as a potential prebiotic starting material for nucleobase synthesis. In the presence of catalysts (for example, TiO_2) and with moderate heating, formamide can pass surface energy barriers, yielding a complete set of nucleic bases and acyclonucleosides, and favoring both phosphorylations and transphosphorylations necessary for life. In the reaction mechanism, interaction with water seems to be an essential factor for the formamide molecule to function [1, 2, 3, 4, 5]. An important type of nucleoside base synthesis can be supposed to take place under UV light in the formamide reaction with a TiO_2 surface, which would underlie the crucial biological significance of this mineral in making compounds of life. In the dark conditions, the experimental data on formamide adsorption at 300 K over the (001) plane of TiO_2 indicate some amount of unreacted formamide and water among the products (CO, H_2, NH_3, HCN). Thus, some concentration of water and its involvement in formamide–TiO_2 interaction was experimentally shown to be an important constituent. Besides nucleoside synthesis, formamide actively influnces *in situ* the hybridization of nucleotide, DNA, and RNA molecules [6, 7, 8, 1, 2, 3, 4, 5].

Formamide (CH_3NO, hereinafter also as FM), also known as methanamide, is an amide derived from formic acid. It is a clear liquid which is miscible with water and has an ammonia-like odor. When heated strongly, FM decomposes to hydrogen cyanide (HCN) and water vapor. As a constituent, FM is also used for the cryopreservation of tissues and organs. In relation to the DNA and RNA molecules, some important formamide's properties can be outlined as follows. It stabilises RNA in gel electrophoresis by deionizing RNA; in capillary electrophoresis, it is used for stabilizing single strands of denatured DNA. FM lowers the melting point of nucleic acids, so the strands separate more readily [6]. Theoretical investigations of the chemical transformations of FM, a molecule of prebiotic interest as a precursor for biomolecules, are conducted concerning the formation of small molecules including CO, NH_3, H_2O, HCN, HNC, H_2, HNCO, and HOCN [9]. The solvent, dehydration, adsorption, and decomposition properties of formamide on various surfaces were intensively studied in recent experimental papers [10, 11, 12, 1, 2, 13]. Some of these experimental results are summarized and descripted in [14]. In [15], it is shown for the first time that guanine, adenine, and hypoxanthine can be produced from formamide in a single model prebiotic reaction at lower temperatures than previously reported if FM is subjected to UV irradiation during heating; this observation relaxes the requirements for prebiotic purine nucleobase formation.

The aim of this work is to simulate the formamide–water (FM–W) and ethanol–water (ETH–W) interaction process on the TiO_2 (anatase) and Pt (platinum) surfaces and to estimate the possible activation energy barrier with the MD method. Molecular modeling was performed for the temperature range of $T = 250$ K to $T = 400$ K, and the diffusion and density profiles were established for the formamide (FM), ethanol (ETH) and water (W) molecules on the TiO_2 (anatase) and Pt (platinum) surfaces. In this regard, metal surfaces are often used in the synthesis and degradation of oxygen-containing compounds, such as alcohols, where carbon–carbon (C–C) and carbon–oxygen (C–O) bond formation and breakage are the elementary steps in this type of process, and the metal surface plays the primary role in the efficiency and selectivity of these steps [22]. Few researches were performed on ethanol–water (ETH–W) surface.

In Ref. [16] studied the structure behavior and diffusion properties for an ETH–W solution and investigated the concentration dependence of properties. Wang Yao-Chun et al. used MD simulation to investigate the behavior of pure water molecules, ethanol molecules, and ETH–W mixture with various weight fractions inside Au nanotubes [17]. D.J. Cooke et al. studied the interface between the $\{10\bar{1}4\}$ surface of calcite and pure ethanol, pure water, and 50:50 mixture (by amount) of water and ethanol [18]. Before, in series of work we performed on ETH–W/Pt system for studying the structural and diffusion behavior of both ETH and W molecules on the surface of Pt (111) [19, 20, 21]. Apart from experimental research, recent theoretical and simulation studies were mostly focused on the FM–W reaction process. Several modern theoretical and molecular dynamics (MD) studies of FM–W interactions should be mentioned in this respect [23, 24]. Nevertheless, to our best knowledge, little is known about the FM/TiO_2 surface interaction, and almost nothing has been reported on the FM–W/TiO_2 one. In this paper, using the MD simulation method, we aimed to make the first attempts to investigate the FM/TiO_2 surface interaction mechanism in detail on the atomic/molecular level. In the presence of water (W), we elucidate the structural, diffusional, and molecular concentration distribution effects. We consider water presence to be a stabilizing factor on the FM/TiO_2 surface interaction mechanism.

2. Models and Simulation Methodology

2.1. Initial Configurations, Molecular Field Topologies

We have simulated four liquid systems (pure FM, pure ETH, and FM–W and ETH–W mixtures) interacting with the TiO_2 and Pt surfaces. The configuration snapshots and schematic views of the modelled systems are isllustrated in Figs. 1 and 2, respectively. Both FM–W and ETH–W mixtures contain 50–50% molecules of FM, ETH and W liquids. The pure (FM, ETH) and aqueous (FM–W, ETH–W) densities vary in the interval of $\rho = 0.7 - 1.0$ g/cm^3. For example, the system with pure FM has 2304 molecules, the FM–W system contains 1152 molecules of FM (formamide) and 1152 molecules of W (water). The ETH and ETH–W systems have also possess similar compositions as FM and FM–W ones. The bulk TiO_2 (anatase) phase was defined by the unit cell lattice vectors of following lengths: $a_0 = b_0 = 3.785$ Å and $c_0 = 9.514$ Å. The bulk Pt (platinum) phase was defined by the unit cell lattice vectors of the lengths: $a_0 = b_0 = c_0 = 3.923$ Å. All the lattice parameters of surface were taken from EIM databases supported by the Russian Foundation for Basic Research [25]. As for the adsorbing surfaces, we have composed totally four layers of 420 TiO_2 molecules (5040 = 1680 (Ti) + 3360 (O) atoms) and four layers of 784 Pt_4 molecules (3136 (Pt) atoms), respectively. The corresponding geometry parameters of the simulated systems are shown in Table 1.

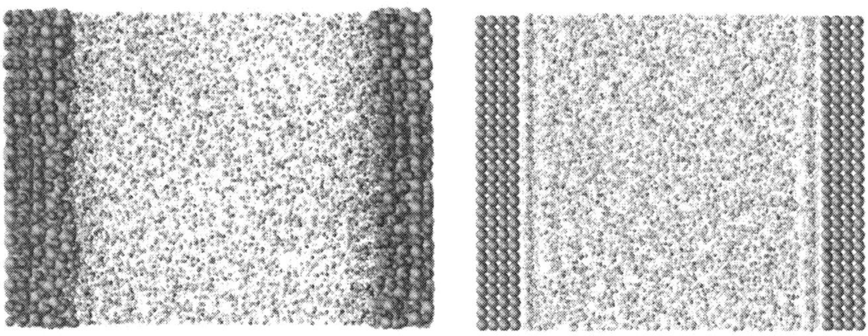

Figure 1. The computer generated snapshots of FM–W mixture relaxed in between two TiO_2 surface walls (left) and ETH–W mixture relaxed inside two Pt surface walls (right) are shown. The green and red large balls indicate Ti and O atoms, gray large balls – Pt atoms, small spheres – the FM–W and ETH–W liquid atoms, respectively.

Table 1. The details of the geometry and molecular compositions of simulated systems

System size	X, Å	Y, Å	Z, Å	ρ, g/cm^3	Ewald parameters (k_1, k_2, k_3)
$(TiO_2)_{5040}(CH_3NO)_{13824}$	56.77	52.99	73.06	1.06	22, 22, 30
$(Pt)_{3136}(CH_3NO)_{13824}$	54.99	54.99	69.22	1.06	22, 22, 28
$(TiO_2)_{5040}(H_2O)_{3456}(CH_3NO)_{6912}$	56.77	52.99	73.06	0.74	22, 22, 30
$(Pt)_{3136}(H_2O)_{3456}(CH_3NO)_{6912}$	54.99	54.99	69.22	0.74	22, 22, 28
$(TiO_2)_{5040}(C_2H_5OH)_{20736}$	56.77	52.99	72.56	1.09	22, 22, 30
$(Pt)_{3136}(C_2H_5OH)_{20736}$	54.92	54.92	69.22	1.09	22, 22, 28
$(TiO_2)_{5040}(H_2O)_{3456}(C_2H_5OH)_{10368}$	56.77	52.99	72.56	0.76	22, 22, 30
$(Pt)_{3136}(H_2O)_{3456}(C_2H_5OH)_{10368}$	54.92	54.92	69.22	0.76	22, 22, 28

For describing of van der Waals interactions we used Lennard-Jones (LJ) potential, which is commonly in use for simulation of liquids. The LJ and electrostatics potentials look like:

$$U = \sum_{ij} \left[4\varepsilon_{ij} \left(\left(\frac{\sigma_{ij}}{r_{ij}} \right)^{12} - \left(\frac{\sigma_{ij}}{r_{ij}} \right)^{6} \right) + \frac{q_i q_j}{r_{ij}} \right] \quad (1)$$

Here $\sigma_{ij}, \varepsilon_{ij}$ are LJ interaction parameters for ith and jth atoms; r_{ij} is interatomic distance; q_i, q_j are partial atomic charges. The LJ, atomic charges and cross-interaction parameters for the FM–W liquid atoms are summarized in Table 5. The cross-section interaction parameters were defined using the Lorentz-Berthelot mixing rule: $\sigma_{ij} = (\sigma_i + \sigma_j)/2, \varepsilon_{ij} = (\varepsilon_i \varepsilon_j)^{1/2}$.

The intramolecular interactions between the liquid molecules were described using the LJ, combined with harmonic, angular and dihedral bonding potentials. For the surface atoms we used the Buckingham [26] and quartic tethering interaction potentials. For the Pt surface the Sutton-Chen potential [27] was used.

The force field models for the TiO_2 (anatase) surface was reported by Kavathekar et al. [28] and Guillot et al. [29]. For the TiO_2 surface models, the potential parameters were developed by Matsui and Akaogi [26]. The pair potential parameters from [30] were used to approximate the Pt–N interaction.

For water, the SPC rigid body model was used [31, 32, 33], and the water bond angles and lengths were not constrained.

2.2. Ensemble, Integration Algorithm, Force Fields

A classical molecular dynamics study was performed using the DL_POLY_2.20 [34] general-purpose code. The NVT ensemble with a Nosé-Hoover thermostat and a Verlet leapfrog scheme were employed. To calculate the long-ranged electrostatic forces, the three dimensional Ewald summation with the automatic optimisation parameter $f = 1.0 \cdot 10^{-4}$ and convergence parameter $\alpha = 0.24375$ Å^{-1} was used. The integration time step of the dynamical equations of motion was 1 fs.

For the FM (formamide), ETH (ethanol) and W (water) molecules, the force field parameters were choosen from the DL_FIELD database [35], which, in their turn, were taken from the CHARMM package database [36].

The detailed potential and force field topology parameters (chemical, angular, and dihedral bonds, the partial atomic charges, etc.) of FM and other components are described below in Tables 4-6.

A parallel Shake algorithm expressed in terms of the replicated data strategy for constraining the rigid and other chemical bonds was used [34].

The MD simulations were realized in the temperature range from $T = 250$ K up to 400 K with a step of $\Delta T = 25$ K.

3. Results and Discussions

We generated the molecular dynamical trajectories and found the relaxed molecular configuration states for the large atomic ensembles that simulate the pure FM and FM–W mixture interacting with TiO_2 (anatase) surface. The FM/TiO_2 and FM–W/TiO_2 interaction processes were performed within the temperature range of T = [250–400] K. Next we repeated the simulations to generate the FM–W interaction on a Pt (platinum) surface, as like as for FM–W/TiO_2 system. Finally, for the sake of comparative analysis, the above procedure have been fulfilled to simulate the ETH–W mixture on the same surfaces (TiO_2, Pt) under similar environmental and thermodynamical parameters.

3.1. Z-density, Dipole Angle Orientations and Diffusion Coefficients

3.1.1. The Density Profiles and Dipole Angle Orientations

The Z-density distributions of the FM and ETH oxygens (Of and Oe atoms), perpendicular to the TiO_2 and Pt surface planes, are shown in Figs. 3(a,b). Fig. 3a compares the Z-densities of pure FM (dashed curve) and FM–W mixture (solid line); Fig. 3b – the Z-densities of pure ETH (dashed curve) and ETH–W mixture (solid line), respectively. Separately, on the same Figs. 3(a,b), the Z-density distributions of water oxygen (OW atom) are presented for the FM–W and ETH–W mixtures, respectively. The comparison of the results in Figs. 3(a,b) indicate that the inclusion of water into consideration has to brought to a substantial enhance of the FM and ETH density distributions in normal direction to the surface plans. On TiO_2 surface the Z-density amplitudes between the pure FM and FM in FM–W mixture have to differ more than twice; on Pt surface the amplitudes have not differ from each other much. The Z-density amplitudes between the pure ETH and ET in ETH–W mixture also differ exactly two times, but it happens on the Pt surface; on TiO_2 surface the amplitudes are close to each other.

Next, one introduced three distances, d_{FM}, d_{ETH} and d_W, to estimate the

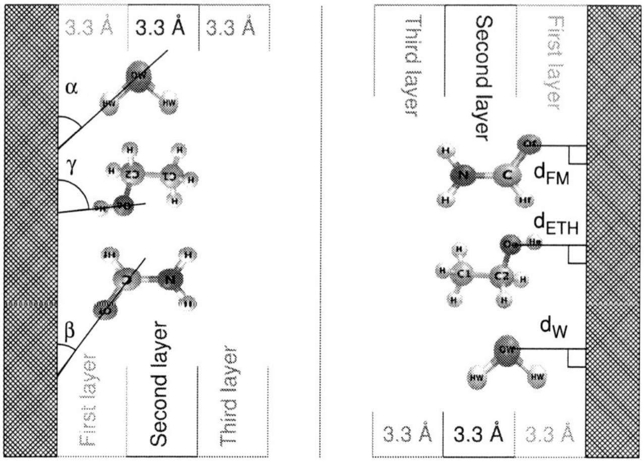

Figure 2. A schematic view of the model systems. The FM molecules are shown as C, N, Of, Hf and H atomic sites; the ETH molecules – as C1, C2, Oe, He and H; the water – as OW and HW, respectively. The α, β and γ show the angles between the surfaces and the liquids OW–HW, C–Of and Oe–He bond distances. The d_W, d_{FM} and d_{ETH} display the minimal contact distances of the liquid oxygens to the surface planes.

minimal contacts between the surface Ti and Pt atoms and the liquid oxygens (Of, Oe and OW). The calculation results for the minimal surface–liquid distances are summarized in Table 2. As it is seen, the minimal contact Ti–Of distance has to be equal to $d_{FM} = 2.863$ Å in absence of water, and equal to $d_{FM} = 2.783$ Å in the presence of water. For water on TiO_2 surface one have obtained that the minimal Ti–OW distance is equal to $d_W = 2.383$ Å.

Table 2. Minimal surface–liquid distances (Å)

Surface	d_{FM}	$d_{FM}(W)$	$d_W(FM)$	d_{ETH}	$d_{ETH}(W)$	$d_W(ETH)$
TiO_2	2.863	2.783	2.383	2.639	2.646	2.389
Pt	2.599	2.517	2.518	2.568	2.585	2.580

Next, for all simulated models, (FM, ETH, FM–W, ETH–W)/(TiO_2, Pt), the angles between the TiO_2 and Pt surface planes and the liquid atomic pair bonds (C–Of and Oe–He) were calculated. The liquid area was divided into three equal

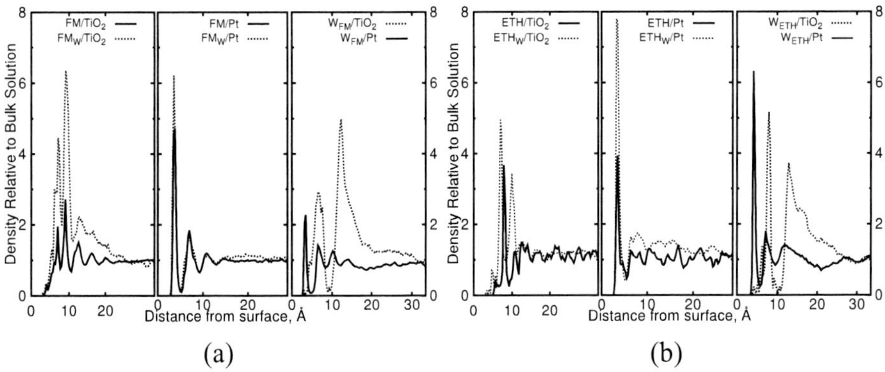

Figure 3. The Z-density distributions of FM, FM–W (a) and ETH, ETH–W (b) solutions on TiO_2 and Pt surfaces.

intervals, each about of 3.3 Å width, and the angles between the liquid bonds and the normal to the surface planes were estimated. Fig. 2 displays the FM–W and ETH–W liquid layers, embedded between two TiO_2 and Pt wall surfaces. The calculation results of the dipole bond distribution are presented in Table 3.

In the first layer, the C–Of bonds of pure formamide on TiO_2 surface are mostly oriented as 80° (91.0%). For the formamide with water the most percentage of the C–Of bonds (65%) in the first layer are oriented under 80° relative to the surface plan. At the same time, the water bonds are being oriented from 110° to 120° (comparing the 4th column with 2nd and 3rd ones in Table 3). The Oe–He bonds of pure ethanol on TiO_2 surface are mostly oriented in 40°–50° (the percentage varies between of 42.0% and 57.0%, respectively). For the ethanol with water the most percentage of the Oe–He bonds (49%) in the first layer are oriented under 40° to the surface. The water bonds of the ETH–W mixture are oriented from 90° to 120° (compare the 7th column with 5th and 6th ones in Table 3). The C–Of bonds of pure FM and FM–W mixture on Pt surface are mostly oriented in 90°–100° (the percentage varies between (48.0%, 49.5%) and (54.5%, 39.0%), respectively). The water OW–HW bonds for both FM–W and ETH–W mixtures are oriented under 90° (compare the 10th and 13th columns with 7th, 8th, 11th and 12th ones in Table 3). The Oe–He bonds of pure ethanol are oriented in 90° (81.5%). For the ETH with water the most percentage of the liquid bonds (45%) are oriented under 70° relative to the surface.

On the second layer, the C–Of dipole angles of both FM and FM–W mix-

Table 3. The dipole angle orientations on the TiO_2 and Pt surface walls (%)

Ang.\	Anatase surface						Platinum surface					
Bond	C-Of	C-Of	OW-HW	Oe-He	Oe-He	OW-HW	C-Of	C-Of	OW-HW	Oe-He	Oe-He	OW-HW
First layer (thickness 3.3 Å, distance from surface 0 Å)												
30					24.0							
40				42.0	49.0							
50				57.0	24.5						0.50	
60		0.50		01.0	02.5						16.5	
70	06.0	34.5							11.5		45.0	05.0
80	91.0	65.0				01.5	01.0	06.5	45.5	02.0	33.5	38.5
90	03.0					13.5	48.0	54.5	38.0	81.5	04.5	53.0
100			06.0			31.5	49.5	39.0	05.0	16.5		03.5
110			49.0			34.0	01.5					
120			39.5			17.0						
130			05.5			02.5						
Second layer (thickness 3.3 Å, distance from surface 3.3 Å)												
50					46.0							
60				0.50	51.5							
70				32.5	02.5		02.0	01.5			02.5	01.0
80			26.0	61.5		18.0	70.5	45.5	13.0		29.0	20.5
90			57.5	05.5		58.0	27.5	50.0	60.5	06.0	52.5	45.0
100	55.0	46.5	16.5			23.0		03.0	26.5	82.0	16.0	29.5
110	45.0	53.5				01.0				12.0		04.0
Third layer (thickness 3.3 Å, distance from surface 6.6 Å)												
70			06.0	37.0		02.5	0.50	01.5			02.5	
80		01.5	57.5	62.0	03.5	49.0	12.0	29.0	23.0	30.5	27.0	14.5
90	0.50	26.5	36.0	01.0	33.5	45.0	76.5	55.0	57.5	69.5	54.5	64.5
100	49.0	58.0	0.50		54.5	03.5	11.0	14.5	18.0		15.5	21.0
110	49.5	14.0			08.5				01.5		0.50	

tures are oriented under 100° and 110° to the TiO_2 surface plane, and under 80° and 90° to the Pt surface (Table 3).

In the third layer, the C–Of bonds of pure formamide are directed under 100° and 110° (the percentage are 49% and 49.5%, respectively); for the formamide with water the C–Of bonds are oriented between 10° (the percentage is 58.0%).

It is worth noting that, the dipole orientations as shown in Table 3 are in a good correlation with the Z-density distributions presented in Figs. 3(a,b) above. From Figs. 3(a,b) one can observe that the Z-density amplitudes of both pure formamide (FM) and formamide with water (FM_W) are located on the same distance from the surface plane (6.6 Å to 9.9 Å) where the second and third layers are positioned. However, the inclusion of water into models substantially influences both the dipole angles and Z-density distributions. The Z-density amplitude of formamide in water is twice larger than for pure formamide, thereby originating from the dipole angle orientations. The comparison of results in Fig. 3(a,b) and Table 3 are straightforward.

3.1.2. Diffusion Coefficients

In this section one investigate the behavior of important self-diffusion coefficient D for the ensembles of the FM, ETH and W molecules. The temperature of the system was varied, and the effect of the temperature changes on the self-diffusion coefficients of the FM, ETH and W molecules were estimated. Fig. 4a presents the values of D for the FM and FM–W mixture, Fig. 4b – for the ETH and ETH–W mixture interacting with the TiO_2 and Pt surface, respectively. The diffusion coefficients decrease with temperature decreasing, which is consistent with the formation of longer H-bonded chains at low temperatures. The presence of the surfaces at low temperatures (275-290 K) have no effect on the value of D for all (FM, ETH and W) liquids.

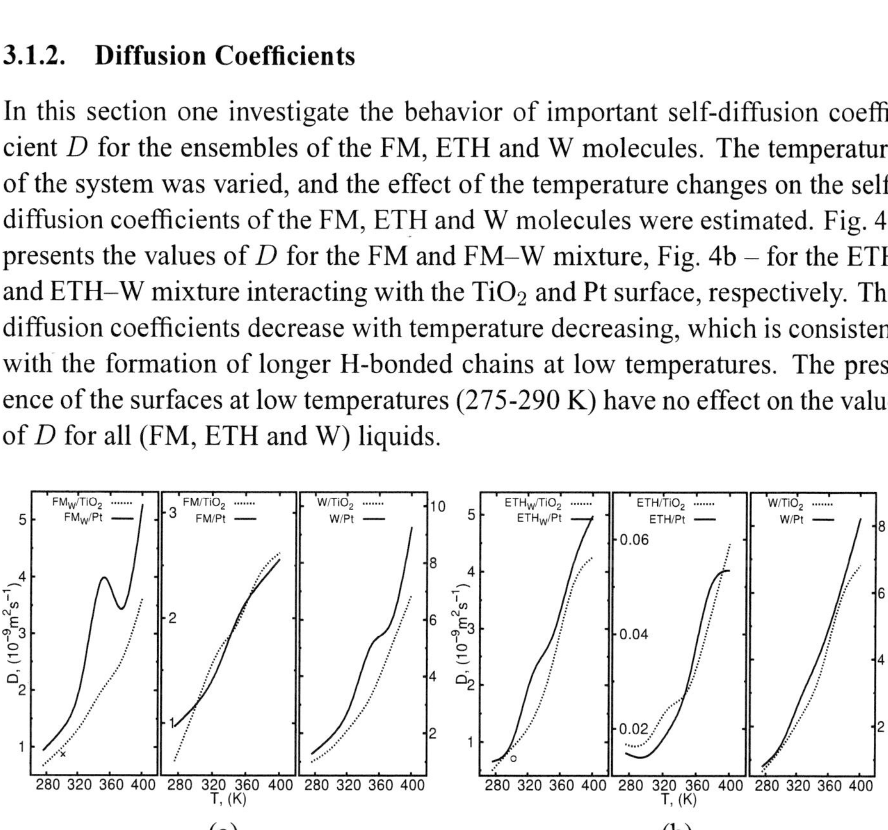

(a) (b)

Figure 4. Self-diffusion coefficients for FM and FM–W mixture and W (a) and for ETH and ETH–W mixture and W (b) as a function of temperature. The × is experimental data taken from Ref. [37, 38]. The ○ is experimental data taken from Ref. [16].

Let us compare the values of the diffusion coefficient D, estimated on TiO_2 (anatase) and Pt surfaces for 50:50% FM–W solution at 300 K, with experimental and other MD simulation results data (by [37], [38], etc). There seen the lack of experimental and simulation data regarding on the (FM–W, ETH–W)/(TiO_2, Pt) systems, so we compared our data with those existed in experiment or other simulations, but without any surfaces. From [37] the diffusion coefficient of the FM molecules was about $0.85 \cdot 10^{-9}$ m^{2}s^{-1} (experimental data for the FM–W mixture (no adsorbing surfaces). From our MD results, we obtain that the value of D_f to be around $0.997 \cdot 10^{-9}$ m^{2}s^{-1} and $1.317 \cdot 10^{-9}$ m^{2}s^{-1} correspondingly to TiO_2 and Pt surfaces. Thus, the presence of the surfaces have to re-estimate the experimental value to about of 17.3% (TiO_2) and 54.9% (Pt), respectively. At the same time, the self-diffusion coefficient of water (in FM–W) on TiO_2 and Pt surfaces have to around of $1.474 \cdot 10^{-9}$ m^{2}s^{-1} and $1.889 \cdot 10^{-9}$ m^{2}s^{-1}, respectively. The latter D values are also higher than the data obtained in [38]: $1.35 \cdot 10^{-9}$ m^{2}s^{-1} (without surface).

The diffusion coefficientestimated for a 50:50% ETH–W solution at 300 K on TiO_2 (anatase) and Pt surfaces were also compared with the experimental and simulation data (by [16], etc). The diffusion coefficient of the ETH molecules from [16] was about $0.7 \cdot 10^{-9}$ m^{2}s^{-1} (experimental data for the ETH–W mixture but without any surfaces). From our results, we obtain that the D_e values to be around $0.901 \cdot 10^{-9}$ m^{2}s^{-1} and $1.01 \cdot 10^{-9}$ m^{2}s^{-1} correspondingly to TiO_2 and Pt surfaces. So far, the presence of the adsorbing surfaces have to re-estimate the existing experimental results to about 28.7% and 44.3% corrections. At the same time, for water (in ETH–W) the self-diffusion coefficient are about $1.348 \cdot 10^{-9}$ m^{2}s^{-1} and $1.471 \cdot 10^{-9}$ m^{2}s^{-1} on TiO_2 and Pt surfaces, which are also higher than the experimental $0.88 \cdot 10^{-9}$ m^{2}s^{-1} ones.

3.2. Peculiarities of Structure Formations

Hereinafter, Hf, HW, Of, and OW denote, respectively, the hydrogen atoms of formamide and water and oxygen atoms of formamide and water.

3.2.1. Liquid–Surface Ordering

The structure of liquids generally is expressed in terms of the radial distribution functions (RDFs) $g(r)$. The most structured and interesting $g(r)$ functions for the liquid like a FM–W solution correspond to the oxygen–oxygen and oxygen–hydrogen bonding. Figs. 5 show the structural arrangement of FM–W system on TiO_2 and Pt surfaces. The Figs. 5 demonstrate the peculiarities of the RDF graphs of the Of–Of, OW–OW, Of–HW, and OW–Hf atomic pairs in

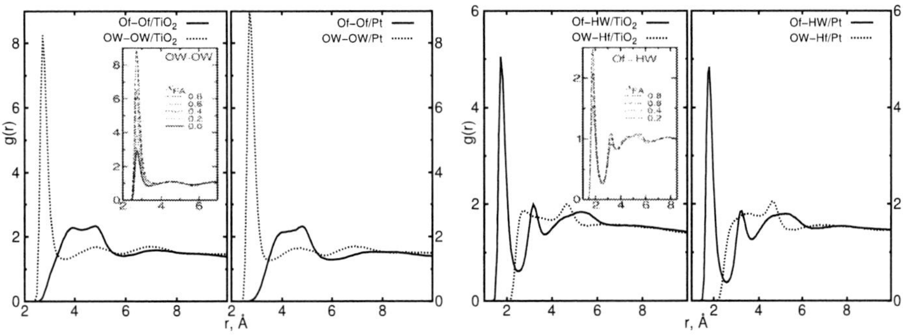

(a) (b)

Figure 5. The FM–FM, W–W and FM–W radial distribution functions (RDFs) $g(r)$ are shown. The $g(r)$ represent the Of–Of, OW–OW (a), Of–HW and OW–Hf (b) atomic pair ordering. Subpictures show the RDFs for the OW–OW and Of–HW atomic pair interactions in the absence of surface taken from Ref. [38].

a formamide–water mixture. It is seen that the behavior of the RDF for water oxygen (OW–OW) is more ordered than that of (Of–Of) one. The first RDF g[OW–OW] peak is located at 3 Å from the zero point, which is 1.5 Å shorter than the g[Of–Of] one. At the same time, the Of–HW interaction looks much stronger than that of the OW–Hf atomic pair. The amplitude of the first g[Of–HW] RDF peak is located at 2 Å and it is twice higher than for g[OW–Hf]. The peak of the g[OW–Hf] is located at 2.8 Å.

We also built the RDF $g(r)$ functions for a ETH–W liquid like a FM–W one with regard to the oxygen–oxygen and oxygen–hydrogen bonding. Hereinafter, He, HW, Oe, and OW denote, respectively, the hydrogen atoms of ethanol and water; and oxygen atoms of ethanol and water. Figs. 6 show the ETH–W/(TiO_2, Pt) structural arrangement, the peculiarities of the RDF graphs of the Oe–Oe, OW–OW, Oe–HW, and OW–He atomic pairs in the ethanol–water mixture. It is seen that the behavior of the RDF curves for water oxygens (OW–OW) is same as for ethanol oxygens (Oe–Oe) one. However, the OW–OW pair shows more stronger ordering, as the amplitudes of the g[OW–OW] are much higher in comparison to g[Oe–Oe] ones (Fig. 6a). The first RDF g[OW–OW] and g[Oe–Oe] peaks are located at 3 Å from the zero point. From Figs. 6b for the RDF (OW–He) and (Oe–HW) one conclude that the Oe–HW interaction has a similar behavior as for OW–He atomic pair. The amplitude of the first RDF g[Oe–HW] peak is located at 2 Å.

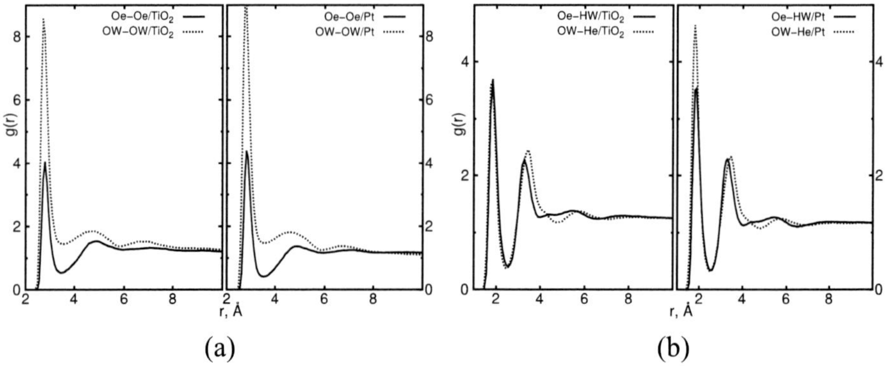

Figure 6. The ETH–ETH, W–W and ETH–W radial distribution functions (RDFs) $g(r)$ are shown. The $g(r)$ represent the Oe–Oe, OW–OW (a), Oe–HW and OW–He (b) atomic pair ordering.

3.2.2. Comparison with Experiment (FM–W System)

As it was pointed out, there is a lack of experimental and simulation data for FM–W/TiO_2, so one compared our results with some experimental and other MD simulations, reported for the absence of the TiO_2 or Pt surface. In Figs. 5 the subpictures are taken from [38], where the RDFs for OW–OW and Of–HW pair interactions of the FM–W mixture were reported at the absence of surface. The comparision show that the height of RDF picks are agreement with accuracy of 20%. It should be mentioned that in our FM–W system the FM ammount has a $\chi \approx 0.6$ mole fraction. Thus, the inclusion of TiO_2 (anatase) surface from our present study seems to essentially re-estimate the structure formation in FM–W system.

Conclusion

Based on the molecular dynamics method and using the DL_POLY_2.20 code, we simulated the FM–W interaction process on a TiO_2 (anatase) surface. In the literature, we have noted the lack of experimental and MD simulation data on the FM–W/TiO_2 system. Our results were compared with the experimental and other MD simulation ones, which were performed in the absence of the surface. Our study shows that the inclusion of the TiO_2 (anatase) surface makes it necessary to re-evaluate the structural and diffusion behavior of the FM–W system. The density distribution calculations were performed for the temperature range of $T = 250$ to 400 K. For pure formamide (FM), the density profile of FM/TiO_2 has a weak adsorption layer between two diffuse layers. The inclusion of water in the FM/TiO_2 model substantially changes the density distribution profiles normal to the surface; the amplitude of the density distribution for the FM–W/TiO_2 surface increases. Water strongly influences the FM diffusion capabilities on the surface. At the same time, comparing the structurally important radial distribution functions, we conclude that in the FM–W/TiO_2 interactions the water molecules actviley push formamide ones from the surface to the bulk area. Thus, including water in the system, one can see that it significantly influences both dynamical and structural behavior of FM–W/TiO_2. A MD analysis of the formamide–water/(TiO_2, Pt) and ethanol–water/(TiO_2, Pt) systems has been performed for a temperature range from 250 K to 400 K. A comparative analysis is performed of the FM–W/TiO_2 and FM–W/Pt and ethanol–water/(TiO_2, Pt) interaction. A MD analysis of the formamide–water/(TiO_2, Pt) and ethanol–water/(TiO_2, Pt) systems has been performed for a wide temperature range from $T = 250$ K up to $T = 400$ K. There is a good agreement between the present MD simulation results and the experimental and other simulation data on the structural, density and diffusion properties of the studied systems. The inclusion of the TiO_2 and Pt surfaces to the formamide–water and ethanol–water systems show that their structural and diffusion properties have to be re-evaluated. The reported data could stimulate further experiments in the field and might be useful for application purposes.

Acknowledgments

This work is a part of our collaboration between JINR (Russia), RIKEN (Japan), and Keio University (Japan). The work has been performed within a research collaboration between JINR (Russia) and Keio University (Japan) and was jointly supported by the Japan Society for the Promotion of Science and the Russian Foundation for Basic Research (grant No.13-04-92100). The work has been supported in part by the Grant in Aid for the Global Center of Excellence Program to the Center for Education and Research of Symbiotic, Safe and Secure System Design from Japan's Ministry of Education, Culture, Sport, and Technology. The MD simulations have been performed using computer software, hardware facilities, and cluster machines at the CICC (JINR), RICC (RIKEN), and the Yasuoka Laboratory of Keio University (Japan). The authors would like to specially thank Mr. Sergei Negovelov (JINR) for technical assistance and helpful comments.

The force field potentials and parameters

Table 4. The force field potentials and parameters for TiO_2 and Pt surfaces

Buckingham potential for TiO_2: $A_{ij}\exp(-r_{ij}/\rho_{ij}) - C_{ij}/r_{ij}^6$				
$i-j$	A_{ij} (kcal mol^{-1})	ρ (Å)	C_{ij} (kcal mol^{-1} Å^6)	
Ti–Ti	717647.4	0.154	121.067	
Ti–O	391049.1	0.194	290.331	
O–O	271716.3	0.234	696.888	
Quartic tethering potential parameters for Pt and TiO_2: $kr^2/2 + k'r^3/3 + k''r^4/4$				
k, kcal mol^{-1}/Å^2		k', kcal mol^{-1}/Å^3		k'', kcal mol^{-1}/Å^4
0.4		0.0		0.4
Sutton-Chen potential parameters for Pt: $\varepsilon[\sum_j (a/r_{ij})^N/2 - C\sqrt{\rho_i}]$, $\rho_i = \sum_j (a/r_{ij})^M$				
ε, kcal mol^{-1}	a, Å	N	M	C
0.226	3.92	11.0	7.0	71.336

Table 5. LJ parameters and charges of solution atoms and cross-interaction parameters for liquid-surface

	Ethanol				Formamide			Liquid-Surface		
Atom	ε, kcal/mol	σ, Å	q/e, p.c.	Atom	ε, kcal/mol	σ, Å	q/e, p.c.	$i-j$	ε, kcal/mol	σ, Å
C1	0.0800	3.6705	−0.27	N	0.2000	3.2963	−0.69	Ti-O	7.7253	2.3431
C2	0.0550	3.8754	+0.05	C	0.0700	3.5636	+0.42	Pt-O	0.9200	2.7000
Oe	0.1521	3.1538	−0.66	Of	0.1200	3.0291	−0.51	Ti-C	7.2630	4.1340
He	0.0460	0.4000	+0.43	Hf	0.0220	2.3520	+0.08	Pt-C	0.9400	2.9000
H	0.0220	2.3520	+0.09	H	0.0460	0.4000	+0.35	Ti-N	0.7010	4.1310
OW	0.1521	3.1506	−0.82	HW	0.0460	0.4000	+0.41	Pt-N	1.7687	3.0615

Table 6. The potential force field parameters for formamide and ethanol molecules

Harmonic bond potential: $K(r_{ij} - r_0)^2/2$					
Bond	K, Å·kcal/mol	r_0, Å	Bond	K, kcal mol^{-1}Å	r_0, Å
1–2	960.00	1.00	1–5	222.00	1.52
1–4	860.00	1.36	1–2	309.00	1.11
4–6	1300.00	1.23	5–6	309.00	1.11
4–5	634.26	1.10	5–8	428.00	1.42
1–6	100.00	2.37	8–9	545.00	0.94
1–5	100.00	1.98	1–3	309.00	1.11
1–3	960.00	1.00	1–4	309.00	1.11
			5–7	309.00	1.11

Lennard–Jones bond potential: $4\varepsilon((\sigma/r_{ij})^{12} - (\sigma/r_{ij})^6)$		
Bond	ε, kcal/mol	σ, Å
2–6	0.0743	1.4473
2–5	0.0318	1.3760
3–6	0.0743	1.4473
3–5	0.0318	1.3760

Angular potential: $K\ (\theta_{ijk} - \theta_0)^2/2$					
Group	K, kcal/(mol rad^2)	θ_0, °	Group	K, kcal/mol rad^{-2}	θ_0, °
2–1–3	46.00	120.00	2–1–3	35.50	109.44
2–1–4	100.00	120.00	6–5–7	35.50	120.00
1–4–6	150.00	122.00	2–1–5	34.60	109.46
1–4–5	88.00	111.00	1–5–8	75.70	109.00
6–4–5	88.00	122.00	1–5–6	34.60	108.89
2–1–4	46.00	120.00	5–8–9	57.50	109.50
3–1–4	100.00	120.00	6–5–8	45.90	108.89

Dihedral potentials: $K[1 + \cos(m\varphi_{ijkn} - \delta)]$				$K(\varphi_{ijkn} - \varphi_0)^2/2$		
Group	K, kcal/mol	δ, °	m	Group	K, kcal/mol	φ_0, °
2–1–4–6	1.40	180	2.00	2–1–5–8	0.16	180
2–1–4–5	1.40	180	2.00	2–1–5–6	0.16	-60
3–1–4–6	1.40	180	2.00	6–5–8–9	0.14	-60
3–1–4–5	1.40	180	2.00	1–5–8–9	1.30	180
				2–1–5–7	0.16	60
				3–1–5–6	0.16	180
				3–1–5–7	0.16	-60
				3–1–5–8	0.16	60
				4–1–5–6	0.16	60
				4–1–5–7	0.16	-180
				4–1–5–8	0.16	-60
				7–5–8–9	0.14	60

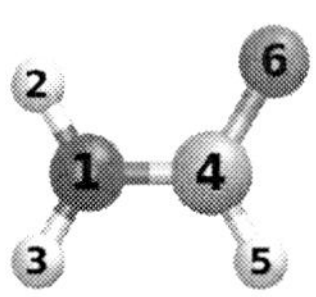

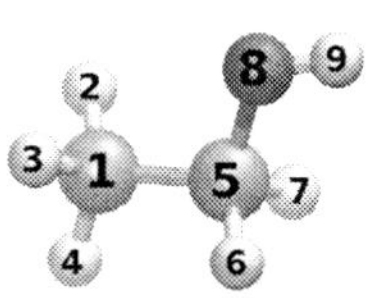

References

[1] Senanayake S.D. and Idriss H. Photocatalysis and the origin of life: Synthesis of nucleoside bases from formamide on TiO_2(001) single surfaces. *Proceedings of the National Academy of Sciences of the United States of America (PNAS)*, 2006, 103 (5), 1194–1198.

[2] Saladino R., Crestini C., Costanzo G., and Di Mauro E. Advances in the prebiotic synthesis of nucleic acids bases: Implications for the origin of life. *Current Organic Chemistry*, 2004, 8(15), 1425–1443.

[3] Ferris J.P., Hill(Jr.) A.R., Liu R., and Orgel L.E. Synthesis of long prebiotic oligomers on mineral surfaces. *Nature*, 1996, 381(6577), 59–61.

[4] Ferris J.P. Catalysis and prebiotic RNA synthesis. *Origins of Life and Evolution of Biospheres*, 1993, 23(5-6), 307–315.

[5] Huber C. and Wächtershäuser G. Peptides by activation of amino acids with CO on (Ni,Fe)s surfaces: Implications for the origin of life. *Science*, 1998, 281(5377), 670–672.

[6] Schoffstall A.M. and Laing E.M. Equilibration of nucleotide derivatives in formamide. *Origins of Life and Evolution of Biospheres*, 1984, 14(1-4), 221–228.

[7] Schoffstall A.M., Barto R.J., and Ramos D.L. Nucleoside and deoxynucleoside phosphorylation in formamide solutions. *Origins of Life and Evolution of Biospheres*, 1982, 12(2), 143–151.

[8] Berndt A., Kosmehl H., Celeda D., and Katenkamp D. Reduced formamide content and hybridization temperature results in increased non-radioactive mRNA *in situ* hybridization signals. *Acta Histochemica*, 1996, 98(1), 79–87.

[9] Nguyen V.S., Abbott H.L., Michele Dawley M., Orlando T.M., Leszczynski J., and Nguyen M.T.. Theoretical study of formamide decomposition pathways. *The Journal of Physical Chemistry A*, 2011, 115(5), 841–851.

[10] Kamineni V.K., Lvov Y.M., and Dobbins T.A. Layer-by-layer nanoassembly of polyelectrolytes using formamide as the working medium. *Langmuir*, 2007, 23(14), 7423–7427.

[11] Parmeter J.E., Schwalke U., and Weinberg W.H. Interaction of formamide with the ru(001) surface. *J. Am. Chem. Soc.*, 1988, 110(1), 53–62.

[12] Muir J.M.R. and Idriss H. Formamide reactions on rutile $TiO_2(011)$ surface. *Surface Science*, 2009, 603(19), 2986–2990.

[13] Saladino R., Ciambecchini U., Crestini C., Costanzo G, Negri R., and Di Mauro E. One-pot TiO_2-catalyzed synthesis of nucleic bases and acyclonucleosides from formamide: Implications for the origin of life. *ChemBioChem*, 2003, 4(6), 514–521.

[14] Saladino R., Crestini C., Ciciriello F., Costanzo G., and DiMauro E.. Formamide chemistry and the origin of informational polymers. *Chemistry & Biodiversity*, 2007, 4(4), 694–720.

[15] BarksH.L., BuckleyR., Grieves G.A., Di Mauro E., Hud N.V., and Orlando T.M. Guanine, adenine, and hypoxanthine production in UV-irradiated formamide solutions: Relaxation of the requirements for prebiotic purine nucleobase formation. *ChemBioChem*, 2010, 11(9), 1240–1243.

[16] Zhang C. and Yang X. Molecular dynamics simulation of ethanol/water mixtures for structure and diffusion properties. *Fluid Phase Equilibria*, 2005, 231(1), 1–10.

[17] Wang Y.-C., Chen C., and Ju S.-P. Adsorption mechanism and dynamic behavior of water and ethanol molecules inside Au nanotubes. *Chinese Journal of Catalysis*, 2008, 29(11), 1099–1106.

[18] Cooke D.J., Gray R.J., Sand K.K., Stipp S.L.S., and Elliott J.A. Interaction of ethanol and water with the $\{10\bar{1}4\}$ surface of calcite. *Langmuir*, 2010, 26(18), 14520–14529.

[19] Kholmurodov K., Dushanov E., Yasuoka K., Khalil H, Galal A., Ahmed S., Sweilam N., and Moharram H. Molecular dynamics simulation of the interaction of ethanol-water mixture with a Pt surface. *Natural Science*, 2011, 3(12), 1011–1021.

[20] Kholmurodov K., Dushanov E., Yasuoka K., Khalil H., Galal A., Ahmed S., Sweilam N., and Moharram H. Molecular dynamics study of ethanol solvated by water on the Pt (111) surface. *Chemical Physics*, 2012, 402(0), 41–47.

[21] Kholmurodov K., Dushanov E., Yasuoka K., Hassan H., Galal A., and Sweilam N.. Structural and diffusional study of pure ethanol and water on Pt(III) surface using molecular dynamics simulation. *European Chemical Bulletin*, 2013, 2(5), 247–254.

[22] Sachtler W.M.H. and Ichikawa M. Catalytic site requirements for elementary steps in syngas conversion to oxygenates over promoted Rhodium. *The Journal of Physical Chemistry*, 1986, 90(20), 4752–4758.

[23] Chalmet S. and Ruiz-López M.S. Molecular dynamics simulation of formamide in water using density functional theory and classical potentials. *Journal of Chemical Physics*, 1999, 111(3), 1117.

[24] Pomata M.H.H, Laria D., Skaf M.S., and Elola M.D. Molecular dynamics simulations of AOT-water/formamide reverse micelles: Structural and dynamical properties. *Journal of Chemical Physics*, 2008, 129(24), 244503.

[25] WWW-MINCRYST. Crystallographic and crystallochemical database for minerals and their structural analogues. 2012. Available from `http://database.iem.ac.ru/mincryst/`

[26] Matsui M. and Akaogi M. Molecular dynamics simulation of the structural and physical properties of the four polymorphs of TiO_2. *Molecular Simulation*, 1991, 6(4-6), 239–244.

[27] Sutton A.P. and Chen J. Long-range Finnis-Sinclair potentials. *Philosophical Magazine Letters*, 1990, 61(3), 139–146.

[28] Kavathekar R.S., Dev P., English N.J., and MacElroy J.M.D. Molecular dynamics study of water in contact with the TiO_2 rutile-110, 100, 101, 001 and anatase-101, 001 surface. *Molecular Physics*, 2011, 109(13), 1649–1656.

[29] Guillot B. and Sator N. A computer simulation study of natural silicate melts. part i: Low pressure properties. *Geochimica et Cosmochimica Acta*, 2007, 71(5), 1249–1265.

[30] Li Zh. and Drazer G. Fluid enhancement of particle transport in nanochannels. *Physics of Fluids*, 2006, 18(11), 117102.

[31] Berendsen H.J.C, Grigera J.R., and Straatsma T.P. The missing term in effective pair potentials. *J. Phys. Chem.*, 1987, 91, 6269–6271.

[32] Robinson G.W., Singh S., Zhu S.-B., and Evans M.W. Water in biology, chemistry and physics: Experimental overviews and computational methodologies. *World Scientific Series in Contemporary Chemical Physics*, 1996, 9, 528.

[33] Kusalik P.G. and Svishchev I.M. The spatial structure in liquid water. *Science*, 1994, 265, 1219–1221.

[34] Forester T.R. and Smith W. Dl_POLY_2.0: A general-purpose parallel molecular dynamics simulation package. *Journal of Molecular Graphics*, 1996, 14(3), 136–141.

[35] Yong C.W. Dl_FIELD — a force field and model development tool for DL_POLY. In Richard Blake, editor, *CSE Frontiers. STFCs Computational Science and Engineering Department (CSED)*. Science and Technology Facilities Council, STFC Daresbury Laboratory, 2010, 2010, 38–40.

[36] B.R. Brooks, C.L. Brooks III, A.D. Mackerell(Jr.), L. Nilsson, R.J. Petrella, B. Roux, Y. Won, G. Archontis, C. Bartels, S. Boresch, A. Caflisch, L. Caves, Q. Cui, A.R. Dinner, M. Feig, S. Fischer, J. Gao, M. Hodoscek, W. Im, K. Kuczera, T. Lazaridis, J. Ma, V. Ovchinnikov, E. Paci, R.W. Pastor, C.B. Post, J.Z. Pu, M. Schaefer, B. Tidor, R.M. Venable, H.L. Woodcock, X. Wu, W. Yang, D.M. York, and M. Karplus. CHARMM: The biomolecular simulation program. *Journal of Computational Chemistry*, 2009, 30(10), 1545–1614.

[37] Finter C.K. and Hertz H.G. NMR relaxation study of I$-$ and Na$+$ solvation structure in formamide (FA) and preferential solvation of these ions in the mixture FA/H_2O. *Zeitschrift für Physikalische Chemie*, 1986, 148(1), 75–96.

[38] Elola M.D. and Ladanyi B.M. Computational study of structural and dynamical properties of formamide-water mixtures. *Journal of Chemical Physics*, 2006, 125(184506), 13.

In: Models in Bioscience and Materials Research ISBN: 978-1-62808-052-0
Editor: Kh. T. Kholmurodov

Chapter 3

SALT EFFECTS ON FOLDING OF VILLIN HEADPIECE SUBDOMAIN HP36 STUDIED BY GENERALIZED-ENSEMBLE SIMULATIONS

***Takao Yoda*[1,*], *Yuji Sugita*[2] *and Yuko Okamoto*[3,4,5,6]**
[1]Department of Computer Bioscience, Nagahama Institute of Bio-Science and Technology, Tamura, Nagahama, Shiga, Japan
[2]RIKEN Advanced Science Institute, 2-1 Hirosawa, Wako, Japan
[3]Department of Physics, Graduate School of Science, Nagoya University, Nagoya, Aichi, Japan
[4]Structural Biology Research Center, Graduate School of Science, Nagoya University, Nagoya, Aichi, Japan
[5]Center for Computational Science, Graduate School of Engineering, Nagoya University, Nagoya, Aichi, Japan
[6]Information Technology Center, Nagoya University Nagoya, Aichi, Japan

Abstract

We report the results of a simulation study on folding of villin headpiece subdomain HP36. We performed multicanonical replica-exchange molecular dynamics simulations of HP36 in the presence and the absence of KCl salt ions in the explicitly represented water solvent. During the production runs, each of which is shorter than the experimental folding

[*]E-mail address: t_yoda@nagahama-i-bio.ac.jp

time, we successfully observed folding events of HP36 to its native-like conformations. We examine salt effects on salt bridges within the protein molecule and discuss about their influence on the hydrophobic-core formation of the globular mini protein.

Keywords: Folding, Villin Headpiece Subdomain, HP36, Multicanonical Replica-Exchange Method, Molecular Dynamics Simulations

1. Introduction

Protein folding is one of the long-standing subjects in biophysics and protein science. It was in 1960's when Anfinsen's famous experiments showed foldability of globular proteins *in vitro* [1]. Today, many researchers in these fields consider that molecular simulation has become a powerful tool for observing protein folding events and other relating conformational changes of protein molecules and their complexes. This is due to developments of efficient simulation techniques, improvements of potential-energy functions (force fields), and expansions of computational power available.

Algorithms for efficient molecular simulation have enabled us to carry out molecular dynamics (MD) simulations of fast-folding small proteins using commonly available PC clusters [2, 3, 4]. Here we report a simulation study on folding of a helical mini protein, chicken villin headpiece subdomain HP36 (HP36), by multicanonical replica-exchange method (MUCAREM) [5, 6]. It is one of the generalized-ensemble algorithms for efficient conformational sampling [7].

Villin is an actin-binding protein (with more than 800 amino-acid residues) that regulates association and dissociation of actin molecules. The C-terminal globular domain of villin with 76 residues is called "headpiece domain". McKnight et al. found that a 35-residue fragment corresponding to the C-terminal half of the headpiece domain cooperatively refolds to its stable native structure with three α-helices [8]. This 35-residue fragment is called as HP35. HP36 is a 36-residue polypeptide that has the same amino-acid sequence as HP35 except for an additional N-terminal methionine residue. While HP36 constitutes about a half of the headpiece domain, it has a globular shape with an isolated hydrophobic core that largely stabilizes the native state by hydrophobic effects [9]. Thus, HP36 has several important common elements and characteristics of globular proteins: secondary structures, hydrophobic core, and cooperative folding to the native structure. Due to these features and the small size, HP36 is an attractive target for protein folding simulation studies.

We previously reported the results of a simulation study on folding of HP36 in explicitly represented water solvent [4]. The initial structure of the protein was a fully extended conformation so that information derived from the native structure in protein data bank (PDB) was not used. We adopted MUCAREM for efficient conformational sampling. Folding events to its native-like conformation with backbone root mean square deviation (RMSD) equal to or smaller than 3.0 Å were observed several times during the production runs, of which the total simulation length was about 2 μs that was shorter than the experimental folding time (5 - 10 μs) [10, 11, 12, 13]. The minimum RMSD of backbone N, C^{α}, and C atoms from the native structure (PDB ID: 1YRF) was 1.1 Å.

Here we focus on the effects of KCl salt (dissolved in solvent) on folding events of HP36. We compared folding simulations of HP36 under two different solvent conditions (with and without KCl salt ions in solvent). Our simulation results have shown that adding KCl salt into solvent decreases the number of salt bridges formed within the protein molecule in a later stage of folding. Further analyses suggested that nonnative nonlocal salt bridges formed in a later stage of folding might retard hydrophobic-core dehydration in pure water.

2. Materials and Methods

The amino-acid sequence of HP36 is MLSDEDFKAVFGMTRSAFANLPLWKQQNLKKEKGLF. We at first performed short replica-exchange molecular dynamics (REMD) simulations from a fully extended conformation of HP36 in the absence of explicitly represented solvent molecules. An "unfolded" conformation without any helices formed was arbitrarily chosen from a high-temperature replica. These steps were followed because the fully extended conformation requires too many water molecules to solvate in a water sphere. In this study, we compared results of folding simulation of HP36 under 0.2 M KCl solution with those in pure water. HP36 was solvated by these solution. To keep the protein molecule solvated, a spherical boundary condition with radius = 30 Å was used. The number of atoms in the protein molecule was 596, and the numbers of atoms in the simulated systems were 11135 and 11102 in the absence and the presence of ions dissolved in the solvent, respectively. CHARMM22 force field with CMAP correction for backbone torsion angles was used for the protein molecule, and TIP3P model was used for water [14, 15, 16]. The single time step for all MD simulations was 1.0 fs. Electrostatic interaction was calculated without cutoff by cell multipole method, in all simulations with explicitly represented solvent. The computer code developed in Refs. [5, 6, 17, 18], which is based on version 2 of PRESTO [19], was used after modification for calculation with CHARMM + CMAP force field.

We adopted multicanonical replica-exchange method (MUCAREM) for efficient conformational sampling [6]. The numbers of replicas were 8 (for HP36 in pure water) and 32 (for HP36 in 0.2 M KCl). See refs 5 and 6 for detailed explanation of REM and MUCAREM, respectively. The total MD lengths of the production runs were 2.28 μs (for HP36 in pure water) and 2.05 μs (for HP36 in 0.2 M KCl). We utilized weighted histogram analysis method (WHAM) for analyses of simulation data [20, 21]. Further detailed explanations of the simulation procedure and analyses are given in ref. 4 for simulations of HP36 in pure water. Those for HP36 in 0.2 M KCl, which closely resemble those for HP36 in pure water, will be given elsewhere [22].

3. Results and Discussion

3.1. Folding Scenario of HP36 in 0.2 M KCl at Room Temperature

In the MUCAREM simulation of HP36 in 0.2 M KCl, we again observed folding events of HP36 to its native-like conformations. The minimum backbone RMSD value from the native structure (with respect to N, C^α, and C atoms in non-terminal 34 residues) was 1.25 Å (See Fig. 1).

In this study, we adopted the number of inter-side-chain nonlocal native contacts formed (N_n) as a "reaction coordinate" for analyses. Here, "non-local" means that the contacting two residues are not in the same secondary-structural element and not next to each other. The larger N_n is, the more native-like the HP36 conformation is. The N_n value for the X-ray structure is 18.

Figure 2(a) shows the backbone RMSD of four 15-residue parts in HP36 from the X-ray structure as functions of N'_n. Here, N'_n is the integer part of $N_n/2$. It is clearly seen that two segments near C-terminus of HP36 acquire native-like backbone shape earlier than segments near N-terminus. This result is similar to the previous observation for HP36 in pure water [4].

Because the phenylalanine residues in the hydrophobic core are located in the N-terminal half of HP36 (Phe7, Phe11, and Phe18), the late construction of the backbone structure at this part naturally results in the late side-chain packing at hydrophobic core, as shown in Fig. 2(b). In Fig. 2(b), heavy-atoms' RMSD

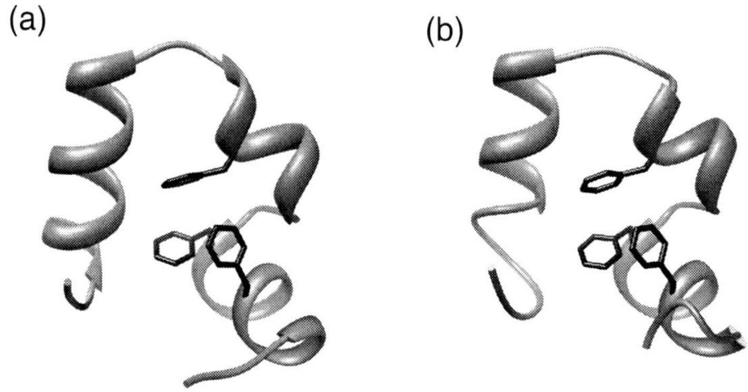

Figure 1. (a) The X-ray structure of HP36 (PDB ID : 1YRF) and (b) the minimum RMSD conformation observed in the MUCAREM simulation of HP36 in 0.2 M KCl. The Phe side chains at the hydrophobic core are shown in the stick representation.

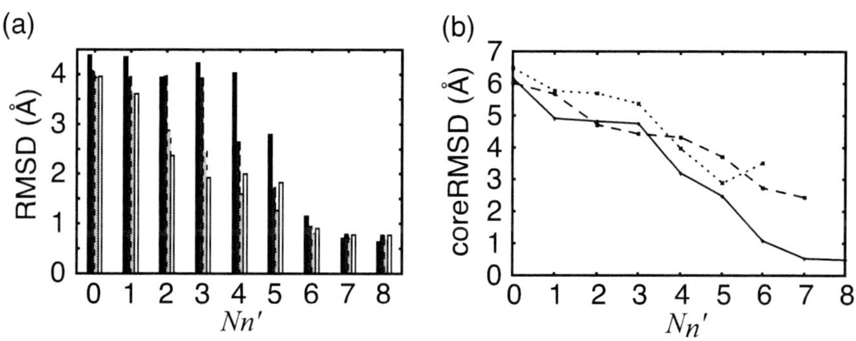

Figure 2. (a) Canonical expectation values of backbone (N, C^{α}, and C) RMSDs for the 15-residue parts of HP36 in 0.2 M KCl at 300 K that were evaluated by WHAM for residues 4-18 (black bars), 9-23 (dark gray bars), 14-28 (gray bars), and 19-33 (open bars). (b) Canonical expectation values of RMSD of heavy atoms in the three Phe side chains at the hydrophobic core of HP36 (coreRMSD) that were evaluated by WHAM. Results for HP36 in 0.2 M KCl at 300 K (solid line) and those for HP36 in water at 350 K (dashed line) and at 300 K (dotted line) are shown in (b). In (a) and (b), values were evaluated as functions of N_n', where N_n' is defined as the integer part of $N_n/2$.

of the side chains of the three Phe at the hydrophobic core is shown as a function of N'_n. The average and the standard deviation values of this quantity in a 115-ns NVT MD of HP36 in pure water at 300 K that was started from the native structure are 0.88 Å and 0.33 Å, respectively. Therefore, Fig. 2(b) suggests that the hydrophobic core does not form native-like conformations until a late stage of folding in which N'_n value is larger than 5, in both conditions.

3.2. Differences in HP36 Folding in the Two Simulation Conditions

Figure 2(b) also shows that the hydrophobic-core packing was more native-like, on the average, in 0.2 M KCl than in pure water, especially in the final stages of folding ($N'_n \geq 6$). We consider that some nonnative salt bridges that are formed at these conformations in pure water distort the hydrophobic-core side-chain packing (see below).

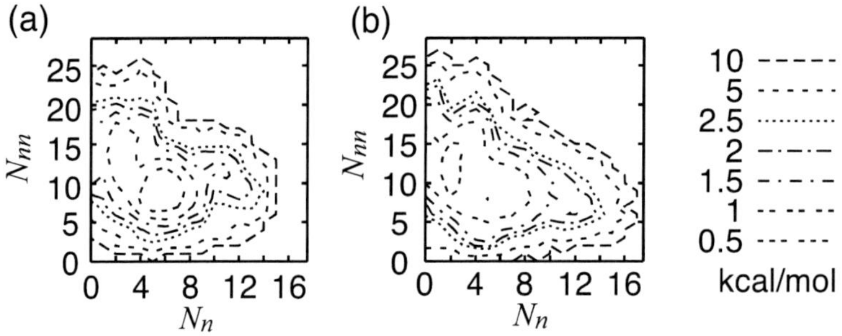

Figure 3. Free-energy maps of HP36 in pure water at 300 K (a) and at 350 K (b) that were evaluated by WHAM. The abscissa and the ordinate axes represent the number of nonlocal inter-side-chain native contacts formed (N_n) and that of non-local inter-side-chain nonnative contacts formed (N_{nn}), respectively. The detailed definitions of N_n and N_{nn} are given in the main text.

Figures 3 and 4 show free-energy maps of HP36 in pure water (Fig. 3) and in 0.2 M KCl (Fig. 4). Here, we use the number of inter-side-chain nonlocal nonnative contacts formed (N_{nn}) as the second reaction coordinate. At 350 K, free energy maps in the two different conditions look similar to each other (see Figs. 3(b) and 4(b)). At 300 K, on the other hand, the populations of the most native-like conformations (the right bottom region in each panel of Figs. 3 and 4) are different (see Figs. 3(a) and 4(a)). N_{nn} value became smaller in

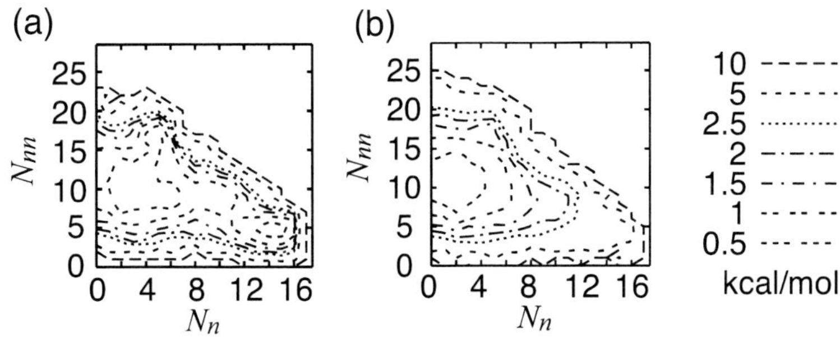

Figure 4. Free-energy maps of HP36 in 0.2 M KCl at 300 K (a) and at 350 K (b) that were evaluated by WHAM. The abscissa and the ordinate axes represent the number of nonlocal inter-side-chain native contacts formed (N_n) and that of non-local inter-side-chain nonnative contacts formed (N_{nn}), respectively. The detailed definitions of N_n and N_{nn} are given in the main text.

0.2 M KCl than in pure water at the latest stage of folding, suggesting some nonnative interactions keeping HP36 molecule from further folding to its fully native structures in pure water.

3.3. Salt-Bridge Analyses

Because the most significant difference between simulation conditions of the two MUCAREM runs is the salt concentration (0 M vs. 0.2 M KCl), we focused our attention on electrostatic interactions, especially on salt-bridging interactions within the protein molecule.

To examine the salt-bridge formation within HP36 that can affect HP36 folding, we analyzed the number of nonnative nonlocal salt bridges formed. Here, "nonlocal" means that the interacting two residues are apart from each other by more than five residues in the primary sequence. We considered that a salt bridge between two charged residues is formed if the distance between at least one pair of atoms in the corresponding charged groups is equal to or less than 3.0 Å. There are seven positively charged residues and five negatively charged residues in the HP36 molecule. In the 35 possible salt-bridging residue pares, 24 pairs are nonlocal ones, among which 21 are nonnative pairs. In the 11 possibly salt-bridging local pairs, there are 9 nonnative ones. It should be noted that the number of nonnative nonlocal salt bridges being able to be formed at

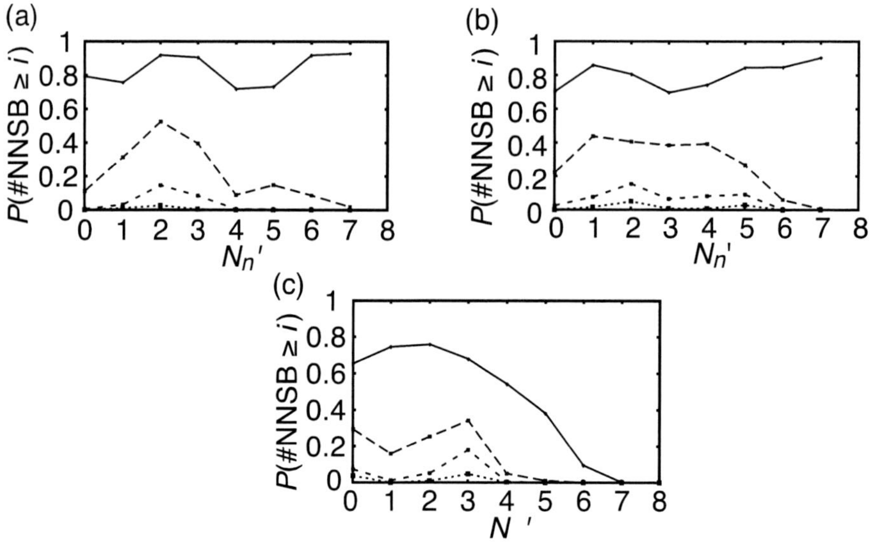

Figure 5. The probability that the number of non-native non-local salt bridges formed (#NNSB) is larger than or equal to the given integer i, where $i = 1$ (solid lines), 2 (dashed lines), 3 (short dashed lines), and 4 (dotted lines). The probability values are evaluated as functions of N'_n, where N'_n is defined as the integer part of $N_n/2$. (a) Those for HP36 in pure water at 300 K evaluated by WHAM. (b) Those for HP36 in pure water at 350 K evaluated by WHAM. (c) Those for HP36 in 0.2 M KCl at 300 K evaluated by WHAM.

one time is much smaller than 21 due to steric hindrance.

Figure 5 shows the fractions of conformations with given numbers of non-native nonlocal salt bridges formed as functions of N'_n. It is clearly seen that the fraction of conformations without such kind of salt bridges formed increases with approaching native state in 0.2 M KCl. In pure water, on the other hand, it turned out that one or more nonnative nonlocal salt bridges kept being formed in most of the native-like conformations with $N'_n \geq 5$.

Protein molecules in the later stages of folding are expected to have more compact shapes than those in the denatured state. Because of this, it seems natural that the chance of nonnative nonlocal salt-bridge formation is as large even in the later stages of folding as in the denatured state. By the addition of ions into solvent, the amount of the nonnative nonlocal salt bridges should become smaller due to electrostatic screening. This can results in the decrease of "misfolding" or "trapping" into conformations that have not only a native-like backbone fold and a lot of inter-side-chain native contacts formed but also some nonnative salt bridges that can slightly distort the side-chain packing in the hydrophobic core.

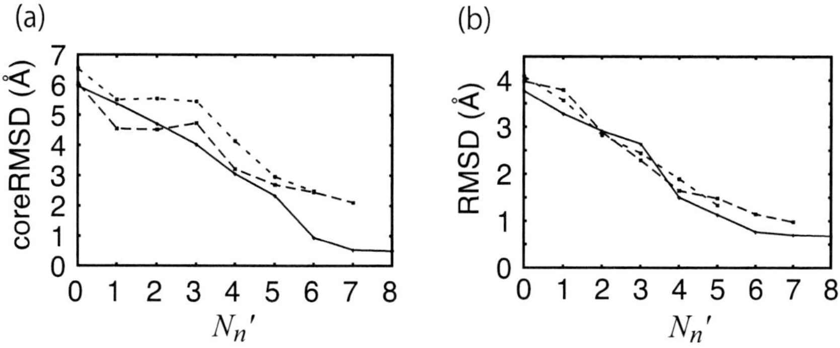

Figure 6. (a) Canonical expectation values of RMSD of heavy atoms in the three Phe side chains at the hydrophobic core (coreRMSD) of HP36 at 300 K in 0.2 M KCl. Those were evaluated by WHAM. (b) Canonical expectation values of backbone (N, C^α, and C) RMSD for residues from 14 to 29 of HP36 in 0.2 M KCl at 300 K evaluated by WHAM. In (a) and (b), values were evaluated as functions of N_n', where N_n' is defined as the integer part of $N_n/2$. Conformations have been divided into three groups according to the number of nonlocal nonnative salt bridges formed (#NNSB): #NNSB = 0 (solid lines), #NNSB = 1 (dashed lines), and #NNSB $\geq$ 2 (short dashed lines).

As for denatured conformations with small N_n' values, the numbers of non-native nonlocal salt bridges formed are not largely different between the two simulation conditions.

To further clarify the effects of nonnative nonlocal salt bridges on hydrophobic-core formation, we evaluated the average RMSD of the three Phe side chains in the hydrophobic core (coreRMSD) as a function of N_n' and the number of nonnative nonlocal salt bridges formed. Fig. 6(a) reveals a correlation between the hydrophobic-core formation and the breakage of nonnative nonlocal salt bridges in the later stages of folding with large N_n' values. This result strongly suggests that nonnative nonlocal salt bridges keep the protein molecule from the fully native hydrophobic-core formation in the later stage of folding.

This salt effect seems to be site-specific. For example, backbone RMSD of residues from 14 to 29, which folds earlier than the other parts of HP36 (see Fig. 2(a)), does not significantly depend on the number of nonnative nonlocal salt bridges formed (see Fig. 6(b)).

It is interesting that the curves of coreRMSD for conformations with one or more nonnative nonlocal salt bridges (dotted line and dashed line in Fig. 6(a)) are similar to the curves for HP36 in pure water (dotted line or dashed line in Fig. 2(b)). This is consistent with the idea that HP36 in pure water tends to get trapped in a later stage of folding due to nonnative nonlocal salt-bridge formations.

Conclusion

In this study we studied the effects of KCl dissolved in solvent on the folding of a helical mini protein HP36 by comparing two sets of folding simulations. The concentrations of KCl in the two simulations were 0 M and 0.2 M. In both simulations, that were started from the fully extended conformation of HP36, we successfully observed folding events of HP36 to its native-like conformations. Our data analyses suggested that early events in folding do not significantly depend on the presence or absence of ions in solvent. On the other hand, the fully native hydrophobic-core formation in the latest stage of folding seems to be affected by nonlocal nonnative salt-bridge formations that likely depend on ion concentrations in solvent. This means that ions dissolved in the solvent can alter the folding landscape or "pathways" especially at the vicinity of the native state. If the present finding mechanism is general in most of globular proteins, ions would be considered as a factor that enhances protein folding in physiological conditions. Further examinations should be done by simulation and experiments.

Acknowledgments

T.Y. was supported by Platform for Drug Discovery, Informatics, and Structural Life Science from the Ministry of Education, Culture, Sports, Science and Technology, Japan (MEXT). Y.O. was supported, in part, by Grants-in-Aid for Scientific Research on Innovative Areas ("Fluctuations and Biological Functions") and for the Computational Materials Science Initiative from MEXT, Japan.

References

[1] Anfinsen, C. B., "Principles that govern the folding of protein chains", *Science* 181, 223-230 (1973).

[2] Sugita, Y. and Okamoto, Y., "Molecular mechanism for stabilizing a short helical peptide studied by generalized-ensemble simulations with explicit solvent", *Biophysical Journal* 88, 3180-3190 (2005).

[3] Yoda, T., Sugita, Y., and Okamoto, Y., "Cooperative folding mechanism of a -hairpin peptide studied by a multicanonical replica-exchange molecular dynamics simulation", *Proteins* 66, 846-859 (2007)

[4] Yoda, T., Sugita, Y., and Okamoto, Y., "Hydrophobic core formation and dehydration in protein folding studied by generalized-ensemble simulations", *Biophys. J.* 99, 1637-1644 (2010)

[5] Sugita, Y. and Okamoto, Y., "Replica-exchange molecular dynamics method for protein folding", Chem. Phys. Lett. 314, 141-151 (1999)

[6] Sugita, Y. and Okamoto, Y., "Replica-exchange multicanonical algorithm and multicanonical replica-exchange method for simulating systems with rough energy landscape", *Chem. Phys. Lett.* 329, 261-270 (2000)

[7] Sugita, Y., Mitsutake, A., and Okamoto, Y., "Generalized-ensemble algorithms for protein folding simulations" in *Lecture Notes in Physics,* Rugged Free Energy Landscapes: Common Computational Approaches in Spin Glasses, Structural Glasses and Biological Macromolecules", W. Janke (ed.) (Springer, Berlin, 2008) pp. 369-407.

[8] McKnight CJ, Doering DS, Matsudaira PT, Kim PS (1996) "A thermostable35-residue subdomain within villin headpiece", *J Mol Biol* 260, 126-134.

[9] Frank, B. S., Vardar, D., Buckley, D. A., and McKnight, C. J. 2002. "The role of aromatic residues in the hydrophobic core of the villin headpiece subdomain". *Protein Sci.* 11: 680-687.

[10] Wang, M., Tang, Y., Sato, S., Vugmeyster, L., McKnight, C.J., and Raleigh, D.P., "Dynamic NMR Line-Shape Analysis Demonstrates that

the Villin Headpiece Subdomain Folds on the Microsecond Time Scale", *J. Am. Chem. Soc.* 125, 6032-6033 (2003)

[11] Kubelka, J., Eaton, W.A., and Hofrichter, J., "Experimental tests of villin subdomain folding simulations", *J. Mol. Biol.* 329, 625-630 (2003)

[12] Buscaglia, M., Kubelka, J., Eaton, and W.A., Hofrichter, J., "Determination of ultrafast protein folding rates from loop formation dynamics", *J. Mol. Biol.* 347, 657-664 (2005)

[13] Kubelka, J., Henry, E.R., Cellmer, T., Hofrichter, J., and Eaton, W.A., "Chemical, physical, and theoretical kinetics of an ultrafast folding protein", *Proc. Natl. Acad. Sci. USA.* 105, 18655-18662 (2008)

[14] MacKerell AD, Jr, Bashford D, Bellott M, Dunbrack RL, Jr, Evanseck JD, Field MJ, Fischer S, Gao J, Guo H, Ha S, Joseph-McCarthy D, Kuchnir L, Kuczera K, Lau FTK, Mattos C, Michnick S, Ngo T, Nguyen DT, Prodhom B, Reiher WE, III, Roux B, Schlenkrich M, Smith JC, Stote R, Straub J, Watanabe M, Wiorkiewicz-Kuczera J, Yin D, and Karplus M. "All-atom empirical potential for molecular modeling and dynamics studies of proteins", *J Phys Chem B* 102: 3586-3616 (1998)

[15] MacKerell, A. D., Jr, Feig, M., and Brooks, C. L., III., "Improved treatment of the protein backbone in empirical force fields", *J. Am. Chem. Soc.* 126, 698-699 (2004)

[16] Mackerell, A.D., Jr., Feig, M., and Brooks III, C.L., "Extending the Treatment of Backbone Energetics in Protein Force Fields: Limitations of Gas-Phase Quantum Mechanics in Reproducing Protein Conformational Distributions in Molecular Dynamics Simulations", *J. Comput. Chem.* 25, 1400-1415 (2004)

[17] Sugita, Y. and Kitao, A., "Improved protein free energy calculation by more accurate treatment of nonbonded energy: application to chymotrypsin inhibitor 2, V57A", *Proteins* 30, 388-400 (1998)

[18] Kitao, A., Hayward, S., and Go, N., "Energy landscape of a native protein: jumping-among-minima model", *Proteins* 33, 496-517 (1998)

[19] Morikami, K., Nakai, T., Kidera, A., Saito, M., and Nakamura, H., "PRESTO (protein engineering simulator): a vectorized molecular dynamics program for biopolymers", *Comput. Chem.* 16, 243-248 (1992)

[20] Ferrenberg, A. M. and Swendsen, R. H., "Optimized Monte Carlo data analysis", *Phys. Rev. Lett.* 63, 1195-1198 (1989)

[21] Kumar, S., Bouzida, D., Swendsen, R. H., Kollman, P. A., and Rosenberg, J. M., "The weighted histogram analysis method for free-energy calculations on biomolecules. I. The method", *J. Comput. Chem.* 13, 1011-1021 (1992)

[22] Yoda, T., Sugita, Y., and Okamoto, Y., in preparation (2012)

In: Models in Bioscience and Materials Research ISBN: 978-1-62808-052-0
Editor: Kh. T. Kholmurodov

Chapter 4

Anisotropic Electrostatics Approaches in Description of Halogen Bonding within Intermolecular Potentials

Oleg I. Titov*[*], *Dmitry A. Shulga, Vladimir A. Palyulin and Nikolai S. Zefirov
Department of Chemistry, M.V. Lomonosov Moscow State University, Moscow, Russia

Abstract

Recently halogen bonding (XB) has attracted much attention as a prospective and previously not used pattern of intermolecular interactions. Though several studies reported marked success in applied rational utilization of XB, its widespread use is hampered by lack of fast and reliable description of XB by means of interatomic potentials. Since the usual systems of interest involving XB are generally too high for *ab initio* treatment at reliable levels, the main alternative is development of fast and robust empirical potentials for XB description fitted to reproduce reference *ab initio* calculations.

In this study the two most promising approaches to description of electrostatic part of XB - extra-point charge and multicenter multipolar distribution - are compared in different contexts: (1) in the reference molecular electrostatic potential (MEP) reproduction, (2) in the reference

[*]E-mail address: titovoi@qsar.chem.msu.ru

potential energy surface reproduction (complemented with van der Waals parameters from GAFF and Tripos force fields), and (3) in estimating the interactions of a small molecule with a protein. Additionally numerical and other aspects of the approaches are compared for the prospects of further development. On the basis of this study multicenter multipolar distribution approach appears to be the most promising for empirical description of electrostatic part of XB.

Keywords: Halogen Bonding, Intermolecular Interactions, Molecular Electrostatic Potential

1. Introduction

Intermolecular interactions determine most physicochemical properties of the organic molecules, and, as a consequence, their understanding and relevant description have proven to give significant insights in drug discovery and material science efforts. Recently halogen bonding (XB), defined as favorable interaction of heavy halogen atom (Cl, Br, I) with Lewis bases, was rediscovered. It has attracted much attention [1, 2] for several reasons. First, it appeared to be of widespread applicability with considerable values of interaction energy (comparable to traditional hydrogen bonding energies). Second, by its ability to form directional interactions in hydrophobic environment, it complements [2] identified previously and widely used interaction patterns fruitfully utilized for decades, such as hydrogen bonding, electrostatic interactions, dispersion interactions, hydrophobic interactions, aromatic stacking. Third, it contradicts traditional conception of halogen in a molecule being only a Lewis base.

After XB was discovered in the 50s it was attributed to charge-transfer complexes. Although the nature of XB is still under investigation, the main hypothesis states that electrostatic interaction is the main factor determining XB energetics [3]. Nowadays, along with electrostatic hypothesis, charge-transfer theory is still respected. DFT level studies show that this contribution is very important in formation of XB [4], but since empirical approaches of charge-transfer description are cumbersome and hence are not well studied, by the moment this hypothesis does not seem to be productive in the applied modeling perspective.

In support of the electrostatic hypothesis, the molecular electrostatic potential (MEP), which governs the electrostatic interactions, around heavy halogen atoms in organic molecules is very characteristic and even counterintuitive. Instead of being isotropic and entirely negative it has a positive area located on the extension of C-Hal bond (Fig. 1). This area was called σ-hole [5]. Its formation is described as a result of nucleus exposure due to anisotropic change of electron density caused by σ-bond formation. This effect of σ-hole formation has common significance: it was found in compounds of elements from VIIth, VIth, Vth to IVth groups of the periodic system [6, 7, 8].

Another interesting consequence of the σ-hole formation near halogen atoms is distinguished anisotropy of repulsive part of van der Waals potential [9]. Since the origin of this interaction is repulsion between electrons, this effect is similarly accounted for by the lack of electron density on the extension of σ-bond, meaning that in this area repulsion should be weaker.

Several models for empirical description of XB were reported earlier. In each model classical electrostatic energy term is corrected to describe XB properties. One can distinguish several groups in these models: models incorporating off-center atomic charges [10, 11, 12], models taking advantage of atomic multipole moments [13, 14] and other approaches [9]. Though these approaches gave positive results, none became a common scheme of choice for empirical description of XB in organic molecules, partially due to the tentative nature of the studies and, to a considerable degree, due to lack of sound and versatile comparison of the drawbacks and benefits brought by each approach. Thus, development of fast empirical models capable of reliable description of XB is of crucial significance to progress in better understanding XB and its successful application.

We conducted a systematic study of different approaches to description of MEP anisotropy for a set of halogen-containing organic molecules and Lewis bases, such as introduction of off-atom centered charges and use of distributed atom-centered multipole expansion. Moreover, the emphasis is put on the potential prospects for the approaches to incorporate into existing (bio-)molecular modeling practices. To that end, the quality of the reference *ab initio* potential energy surface (PES) reproduction by means of the force fields complemented with the studied XB electrostatic approaches is scrutinized.

2. Method

2.1. Atomic Point Charges

This group defines a classical means used over decades in describing electrostatic interaction between molecular species within the force field context. Position of each charge center coincides with positions of atoms in a species in question. Atomic charges were used only for reference purposes since their isotropic Coulombic potential is incapable of well describing XB features (known as the fallacy of atomic charges [15]). AM1-BCC [16] charges were derived via AmberTools software kit [17]. MK-ESP- [18] (or RESP- [19]) like charges were derived by in-house software based on the MEP calculated quantum chemically.

2.2. Extra-Point Charges

The first and the most intuitive scheme of introduction of atomic potential anisotropy is placement of off-atom charge centers. This approach is natural for most of modern software and can be easily used in different calculations. In XB case extra-point (EP) charge is located on the elongation of σ-bond C-Hal, has positive value and represents the lack of electron density in this area. This approach provides good description of MEP and gives good quantitative XB energy estimation [12], but it also has several disadvantages.

The position of off-center charge is not well defined from the theoretical point of view. There are two fundamental ways to do this: (1) fit the position to represent pure MEP, or (2) fit it to represent PES by means of the full force field for which electrostatic interactions are being tuned. The second way provides more convenient parameters for modeling purposes but the resulting parameters can not be classified as parameters for pure electrostatic interaction. The latter additionally complicates building empirical methods to describe XB. We used the first way of defining extra-point position because it is more straightforward and results should be more transferable between different empirical potentials.

2.3. Distributed Multipoles

The second approach of creating anisotropic atomic potential is the use of multicenter multipole expansion, which is the most general and natural from the theoretical point of view. Visual inspection of the MEP of a halogen atom in the organic species suggests that atomic quadrupole is needed to represent its anisotropy. We decided to add atomic dipoles and quadrupoles on anisotropic atoms (halogens in XB donors and nitrogen and oxygen in XB acceptors). Additional multipoles on hydrogens and sp^3 carbon atoms do not have much sense since the former has too few electrons and the latter possesses too high tetrahedral symmetry to have a pronounced anisotropy of potential. Moreover, tetrahedral carbon is hidden behind other atoms. The traceless form of quadrupole rep-

SMD values for different molecules are high (table 1). Both the models give significant increase in accuracy. It should be noted the lowest errors of MEP description were achieved within the oach. Moreover, the error values between the structures are much for MDQ compared to EP and AC methods.

scription

final purpose of the XB electrostatic models is to better esti- intermolecule interactions, the reference PES reproduction is a criterion. We calculated molecular mechanics (MM) interac- XB complex of CH_3Br and NH_3 varying distance between the mine atoms. Counterpoise corrected [20] MP2/aug-cc-pVTZ s reference. As for MM energy, different electrostatic and van were tried. We checked classical (AC), EP and MDQ models Van der Waals energy terms were taken from GAFF [27] and ce fields.

entioned that classical procedure of force field construction s potential calibration after selecting an exact form of elec- in any common force field van der Waals interactions are lectrostatics. This means that any widely used force field can der Waals potential "off the shelf" which fitted our studied .

(Fig. 2) that classical atomic charge approach results in no . Models with improved electrostatics perform better. Both ches describe attraction in the region of interest, however s slightly better. The latter is in accordance with slightly n.

d angular dependencies reproduction (2D PES). We per- ergy along the angular and radial coordinates. Methyl- s fixed with bromine in the coordinate origin. Ammonia omine atom in such a way that its lone pair was always e atom (Fig. 3). We used the same reference as in the n above. For MM models GAFF+EP and Tripos+MDQ sen as the most promising (Fig. 4).

odels - EP and MDQ - reproduce angular dependence ingly well. Although as we pointed out above, separate

resentation is chosen to ensure better transferability and interpretability of the parameters fitted. Further, this model will be referred to as MDQ (monopole, dipole and quadrupole).

2.4. Molecule Set

It is known that MEP around halogen atom depends significantly on substituents. To make a molecule set both compact and diverse in MEP property we have chosen substituents substantially differing in electronegativity: we studied hydrides, fluorides, methyl- and thrifluoromethylderivatives of Cl, Br and I. As a model acceptor for XB we studied ammonia as the most simple and pronounced Lewis base.

2.5. *Ab Initio* Calculations

MP2/aug-cc-pVTZ level of theory was used for MEP calculation and geometry optimizations of isolated molecules. For heavy halogen atoms (Br, I) aug-cc-pVTZ-PP basis set was used. The same approach with counterpoise corrections [20] was used for estimation of PES of XB complexes. All *ab initio* MEP calculations and geometry optimization of isolated molecules were done with Firefly 7.1.G program [21]. XB complex optimization and PES scans were made in Molpro v 2010.1 program suite [22]. During the geometry optimization of XB complexes the geometries of the comprising monomers were frozen.

2.6. Molecular Mechanics Calculations

Molecular mechanics PES estimation was performed with the help of in-house software based on OpenBabel C++ library [23, 24].

2.7. Electrostatic Fitting

For the purposes of fitting the electrostatic models parameters the MEP was calculated on the grid built of 5 spherical layers with radii equal to 1.4, 1.6, 1.8, 2.0 and 2.2 atomic van der Waals radii and point density 1.12 Å^{-2}. In-house software was used for fitting parameters for all the models. The fit was performed via SVD analysis as described in [25]. SVD was performed via DGESDD routine from LAPACK package [26]. During the fit process constraints on the total charge and symmetry of equivalent atoms were applied. Moreover, additional constraints on symmetry and orientation of multipole moments according to molecular symmetry were added. Atomic types, multipole axes orientations based on the symmetry of fragments were assigned according to specially derived rules and were implemented, along with atom equivalency detection, with the help of OpenBabel C++ library [23, 24].

3. Results

3.1. MEP Description

The MEP reproduction in different electrostatic approaches is shown in Fig. 1. On the *ab initio* potential plot the σ-hole is well seen (to the right of the bromine atom). Atomic charges are evidently unable to describe such anisotropy.

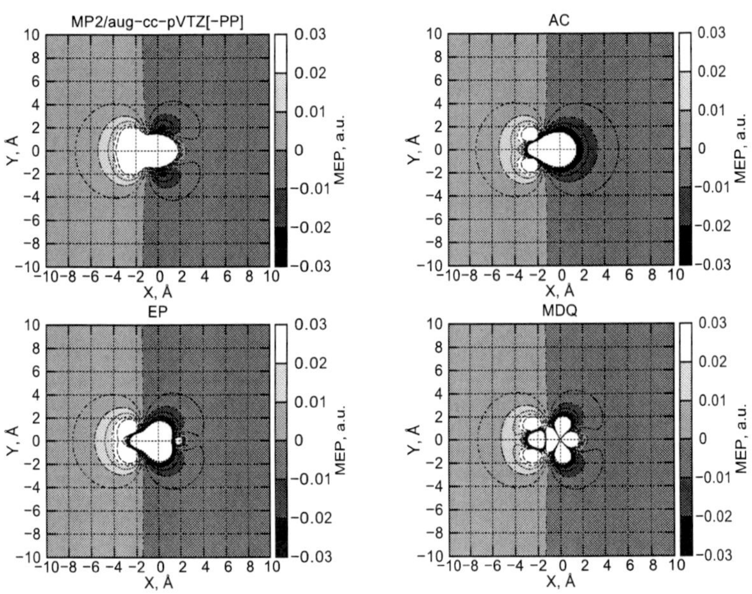

Figure 1. MEP of methylbromide molecule, calculated in the plane containing both bromine (in the coordinates origin) and carbon atoms: *ab initio* calculation (MP2/aug-cc-pVTZ[-PP]); atomic charges (AC); optimal extra-point charges (EP); multicenter multipole expansion (charge, dipole and quadrupole on bromine atom, charges on all other atoms, MDQ).

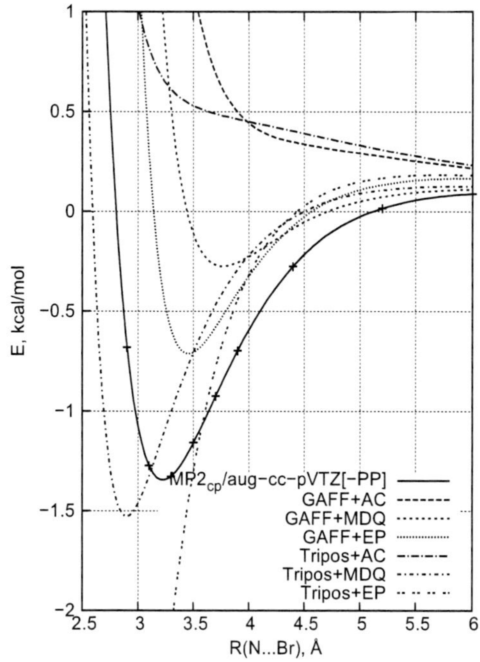

Figure 2. XB PES of molecules CH_3Br and NH_3 calculated by *ab initio* method ($MP2_{cp}$/aug-cc-pVTZ[-PP]) and different combinations of van der Waals potentials (taken from Tripos and GAFF force fields) and electrostatic models: atomic charges (AC), optimal extra-point charges (EP) and multicenter multipole expansion with charge, dipole and quadrupole on bromine atom and charges on all the other atoms (MDQ).

calibration of van der Waals parameters might well distinguish the achieved quality of angular PES reproduction.

3.3. An Applied Case Study

It was very instructive to conduct an experiment to test the models in the settings closest to their intended use: estimation of PES of interaction of a small molecule with a biomolecular system. As a test system the crystallographic structure (PDB ID: 1P5E) of the complex of cyclin-dependent kinase 2 (*phospho*-CDK2) with 4,5,6,7-tetrabrombenzotriazole (TBB) was used. TBB

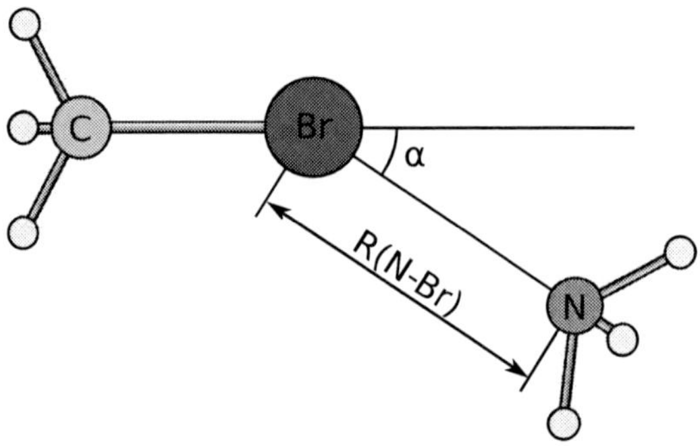

Figure 3. 2D XB PES scan scheme.

is a competitive inhibitor which binds to the ATP binding site. This protein plays an important role in the process of cell division and its inhibitors are believed to be potential anti-cancer drugs. TBB forms several XB bonds in the ligand pocket and this complex seems to be a good test system for XB models. We scanned energy of the ligand leaving complex along the straight line (with frozen geometry of both ligand and protein). Though this trajectory is far from the optimal free energy path, the results are very instructive from the point of view of comparing the electrostatic approaches in the settings close to realistic.

It should be noted that the crystallographic data reveal several water molecules inside the binding pocket connected with both protein and ligand by hydrogen bonds. Usual practice is to remove all water molecules from a protein structure before molecular modeling study of any protein-ligand complex. But in this case we decided to complement the model obtained by this way with the model in which two water molecules in the crystal structure were evidently forming hydrogen bonds with both protein and ligand. These molecules of water can play important role in ligand binding. Thus two separate experiments were conducted: with two water molecules left in the pocket and without any water.

Since the studied system is considered huge for *ab initio* calculations, no reference data are available here. As for MM models we used combination of GAFF van der Waals potential plus classical electrostatics with AM1-BCC charges as commonly used force field for modeling such systems. Another model used GAFF van der Waals with atomic charges fit directly to MEP (AC). For EP electrostatic model we used parameters reported by Ibrahim (2.2 Å distance between extra-point and bromine atom, RESP fit - aimed to reproduce

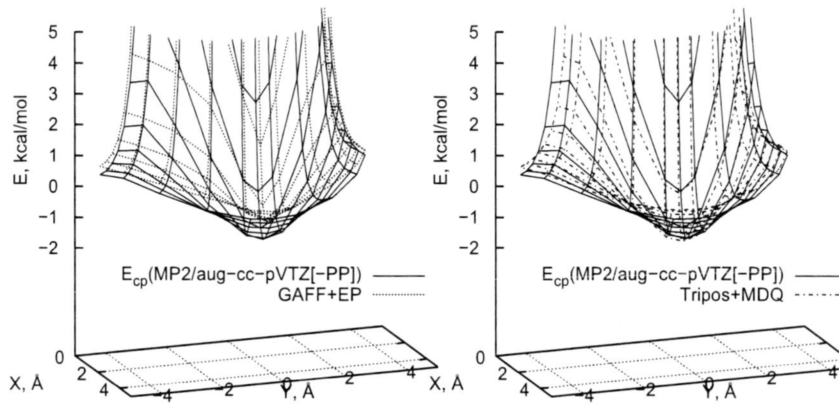

Figure 4. 2D XB PES of molecules CH_3Br and NH_3 calculated by *ab initio* method ($MP2_{cp}$/aug-cc-pVTZ[-PP]) and two molecular mechanics approaches: van der Waals potential taken from GAFF force field and electrostatic model of optimal extra-point charges (left); van der Waals potential taken from Tripos force field and electrostatic model of charge, dipole and quadrupole on bromine atom and charges on all the other atoms (right). Methylbromide molecule is fixed with bromine atom in the origin and carbon lying on the negative X axis, while ammonia is being moved along XY plane with its lone pair (nitrogen) always pointing to the bromine atom.

PES of XB complexes) [12]. The last model uses GAFF van der Waals potential with multipole model with atomic dipole and quadrupole on bromine atoms and atomic charges on the other atoms. We did not used Tripos force field since it gave unreliable results in this system with strong van der Waals repulsion.

It is well seen that without water inclusion classical force field models show repulsion at crystallographic position whereas both enhanced models exhibit strong attraction. When water molecules are included as part of the protein,

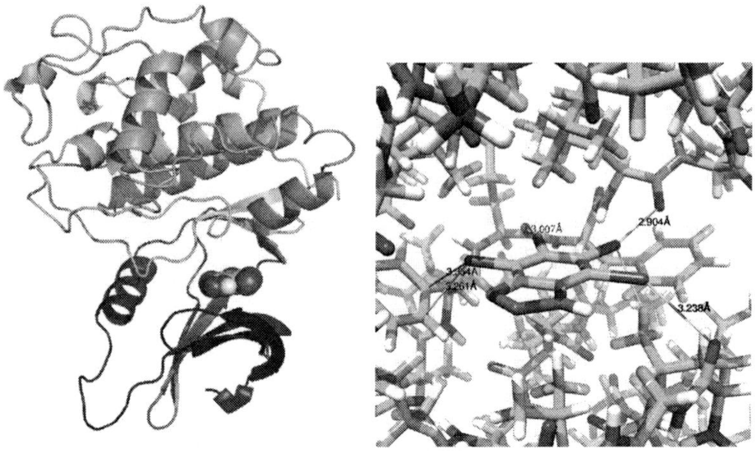

Figure 5. Crystallographic structure of the XB-bonded complex between *phospho*-CDK2 and TBB.

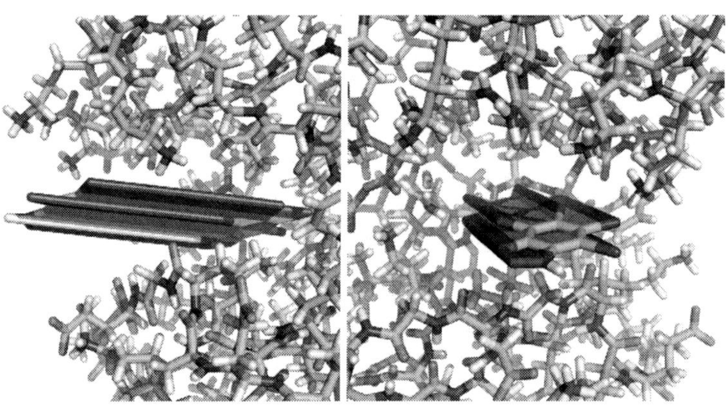

Figure 6. The scanned trajectory of TBB leaving the binding pocket.

all electrostatic energies decrease but still the enhanced models predict stronger attraction than the classical approach does. In both cases (with and without water) difference between enhanced and classical models is near 6 kcal/mol which is comparable with the energy of a hydrogen bond.

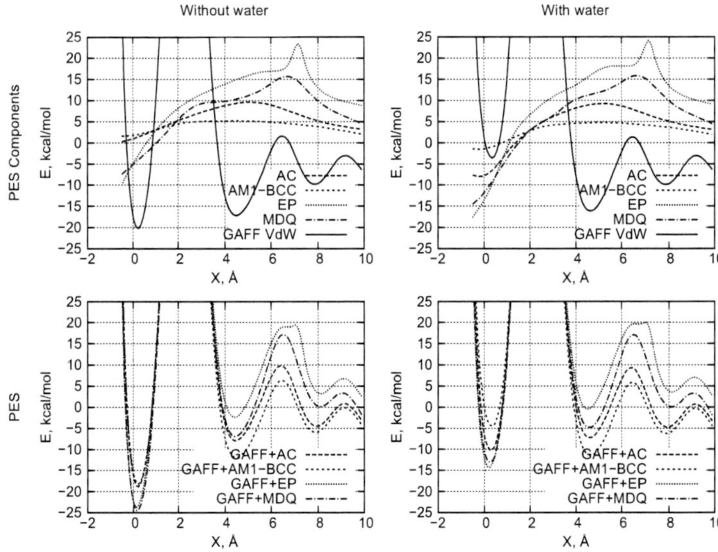

Figure 7. PES of TBB removal from its binding pocket. Crystallographic position is at X = 0 Å.

4. Discussion

There are several important aspects of MEP description approaches that should be taken into consideration during their comparison. It is important for a model to be precise but at the same time it should have as few parameters as possible. Early stages of drug design pipeline demand the existence of fast and efficient ways to obtain model parameters for any molecule, so a route to create such method must exist. To that end, the model and its parameters should be clear and intuitive. The model parameters should also be well defined. Additional consideration is complexity of calculations needed for energy estimation.

4.1. Extra-Point Charges

Introduction of extra-point charge on the C-Hal axis provides means to describe σ-hole MEP anisotropy. As was mentioned above position of such extra-point is not defined theoretically and two means of definition exist: (1) fit to MEP or (2) fit to PES, which do not give the same result. Also the position of extra charge varies much even for the same halogen atom (table 2). It seems that position of extra-point charge is ill-defined in MEP fitting.

With variations of position the values of charges vary greatly (table 2). Moreover, charge values vary in an unpredictable way (e.g. in methylhalogenide series carbon charge shows non-periodic tendency). These complex deviations cannot be easily connected with molecular structure and certainly create difficulties for empirical description.

Another consideration according to extra-point charges is long distance (approx 2.0 Å) between additional charge position and bromine atom. Since usually extra-points do not have any van der Waals potential, this may bring additional artifacts on PES and give numerical errors leading to unreliable results of applied molecular simulations. In fact, we already encountered such artifacts twice when we discussed PES reproduction.

First, when using Tripos force field van der Waals estimation with EP electrostatics, no repulsion is observed between methylbromide and ammonia in the complex (Fig. 2). The optimal position of extra-point for methylbromide molecule is 1.9 Å, at the same time equilibrium distance between bromine and nitrogen atom in the studied complex is 3.2 Å ($MP2_{cp}$/aug-cc-pVTZ). This means that extra-point is only 1.3 Å away from nitrogen atom. This results in extremely favorable interaction between these two charge centers, being of opposite signs, and Tripos potential can not compensate this artificial unbounded attraction.

Second, in the case study of protein-ligand complex a spike in electrostatic EP energy near 7 Å (Fig. 7) is observed. As it can be seen on the trajectory picture (Fig. 6) TBB is moved near a protonated lysine residue. In the closest point one of TBB's bromine is about 2.8 Å from the lysine's nitrogen. Evidently, the extra-point is even closer, which results in high electrostatic repulsion at reasonable interatomic separations.

4.2. Atomic Multipoles

The atomic multipoles model looks more promising (table 3) compared to the extra-point model. The only departure from the periodic dependency is observed in methylderivatives where halogen atoms adopt slightly positive charge. All the other charge values vary according to the concepts of electronegativity, quadrupole moments increase with the increase of atomic number of halogen, etc. This shows good perspectives in building empirical schemes of obtaining interaction potentials approaching *ab initio* quality without the need of *ab initio* calculations, which is extremely desirable for potential application of XB modeling, such as in e.g. drug design pipeline.

Table 2. Optimal extra-point parameters. α and β atoms are defined according to halogen atom

Parameter	H			F			CH_3			CF_3		
	Cl	Br	I	Cl	Br	I	Cl	Br	I	Cl	Br	I
R(EP), Å	0.7	0.8	1.1	1.6	1.7	1.8	2.0	2.0	1.9	1.2	1.2	1.3
q(EP), a. u.	0.381	0.348	0.253	0.071	0.087	0.099	0.021	0.038	0.064	0.117	0.164	0.197
q(X), a. u.	-0.758	-0.663	-0.482	-0.047	-0.002	0.016	-0.198	-0.202	-0.236	-0.263	-0.308	-0.331
q(α), a. u.	0.377	0.315	0.229	-0.024	-0.085	-0.115	-0.209	-0.249	-0.110	0.622	0.630	0.688
q(β), a. u.	-	-	-	-	-	-	0.129	0.137	0.094	-0.159	-0.162	-0.185
RMSD, kcal/mol*e	0.270	0.328	0.778	0.640	0.786	1.518	1.037	1.080	1.507	0.516	0.533	0.725

Table 3. Multipole parameters. α and β atoms are defined according to halogen atom

Parameter	H			F			CH_3			CF_3		
	Cl	Br	I	Cl	Br	I	Cl	Br	I	Cl	Br	I
q(X), a. u.	-0.334	-0.257	-0.132	0.169	0.238	0.349	0.025	0.031	0.052	-0.081	-0.074	-0.051
d(X), a. u.	-0.365	-0.370	-0.235	0.155	0.223	0.495	0.536	0.508	0.533	-0.039	-0.107	-0.155
Q(X), a. u.	0.818	1.205	1.967	1.719	2.472	3.943	2.026	2.584	3.429	1.232	1.629	2.210
q(α), a. u.	0.334	0.257	0.132	-0.169	-0.238	-0.349	-0.497	-0.524	-0.528	0.564	0.558	0.583
q(β), a. u.	-	-	-	-	-	-	0.158	0.164	0.159	-0.161	-0.162	-0.177
RMSD, kcal/mol*e	0.298	0.365	0.782	0.586	0.603	0.862	0.401	0.409	0.689	0.488	0.494	0.683

In contrast to EP model MDQ model shows no side effects on PES reproduction. The chief drawback is a technical one: the majority of modern molecular modeling software do not have interfaces to work with multipoles, so special tricks must be used.

Conclusion

The two most promising approaches to description of anisotropic electrostatic part of halogen bonding - extra-point charge and multicenter multipolar distribution - were compared in different contexts close to their potential application. For the models based on both approaches parameters were fitted to best reproduce the reference *ab initio* molecular electrostatic potential (MEP) around a series of C-Hal containing molecules. Both models qualitatively enhanced MEP description compared to the currently *de facto* default atomic charge model. However careful examination of the achieved quality of both the MEP and PES of interaction reproduction revealed that the multicenter multipolar approach gives more precise results, leads to the proper periodic variation of the fitted parameters and is overall more promising for further development to evolve in a standard part of halogen bonding electrostatic description in intermolecular potentials of broad applicability.

Acknowledgment

The work was supported by Russian Foundation for Basic Research (grant # 11-03-01174-a).

References

[1] Peter Politzer, Pat Lane, Monica C. Concha, Yuguang Ma, and Jane S. Murray. An overview of halogen bonding. *Journal of Molecular Modeling*, 13:305–311, 2007. ISSN 1610-2940. doi: 10.1007/s00894-006-0154-7.

[2] Yunxiang Lu, Yingtao Liu, Zhijian Xu, Haiying Li, Honglai Liu, and Weiliang Zhu. Halogen bonding for rational drug design and new drug discovery. *Expert Opinion on Drug Discovery*, 7(5):375–383, 2012. doi: 10.1517/17460441.2012.678829.

[3] Peter Politzer, Jane S. Murray, and Timothy Clark. Halogen bonding: an electrostatically-driven highly directional noncovalent interaction. *Phys. Chem. Chem. Phys.*, 12:7748–7757, 2010. doi: 10.1039/C004189K.

[4] Marcin Palusiak. On the nature of halogen bond the kohnsham molecular orbital approach. *Journal of Molecular Structure: THEOCHEM*, 945(13): 89 – 92, 2010. ISSN 0166-1280. doi: 10.1016/j.theochem.2010.01.022.

[5] Timothy Clark, Matthias Hennemann, Jane S. Murray, and Peter Politzer. Halogen bonding: the σ-hole. *Journal of Molecular Modeling*, 13:291–296, 2007. ISSN 1610-2940. doi: 10.1007/s00894-006-0130-2.

[6] Jane S. Murray, Pat Lane, Timothy Clark, and Peter Politzer. σ-hole bonding: molecules containing group VI atoms. *Journal of Molecular Modeling*, 13:1033–1038, 2007a. ISSN 1610-2940. doi: 10.1007/s00894-007-0225-4.

[7] Jane S. Murray, Pat Lane, and Peter Politzer. A predicted new type of directional noncovalent interaction. *International Journal of Quantum Chemistry*, 107(12):2286–2292, 2007b. ISSN 1097-461X. doi: 10.1002/qua.21352.

[8] Jane S. Murray, Pat Lane, and Peter Politzer. Expansion of the σ-hole concept. *Journal of Molecular Modeling*, 15:723–729, 2009. ISSN 1610-2940. doi: 10.1007/s00894-008-0386-9.

[9] Megan Carter, Anthony K. Rappé, and P. Shing Ho. Scalable anisotropic shape and electrostatic models for biological bromine halogen bonds. *Journal of Chemical Theory and Computation*, 8(7):2461–2473, 2012. doi: 10.1021/ct3001969.

[10] Zhijian Xu, Zheng Liu, Tong Chen, TianTian Chen, Zhen Wang, Guanghui Tian, Jing Shi, Xuelan Wang, Yunxiang Lu, Xiuhua Yan, Guan Wang, Hualiang Jiang, Kaixian Chen, Shudong Wang, Yechun Xu, Jingshan Shen, and Weiliang Zhu. Utilization of halogen bond in lead optimization: A case study of rational design of potent phosphodiesterase type 5 (pde5) inhibitors. *Journal of Medicinal Chemistry*, 54(15):5607–5611, 2011. doi: 10.1021/jm200644r.

[11] Stefano Rendine, Stefano Pieraccini, Alessandra Forni, and Maurizio Sironi. Halogen bonding in ligand-receptor systems in the framework of classical force fields. *Phys. Chem. Chem. Phys.*, 13:19508–19516, 2011. doi: 10.1039/C1CP22436K.

[12] Mahmoud A. A. Ibrahim. Molecular mechanical study of halogen bonding in drug discovery. *Journal of Computational Chemistry*, 32(12):2564–2574, 2011. ISSN 1096-987X. doi: 10.1002/jcc.21836.

[13] Hajime Torii and Masaki Yoshida. Properties of halogen atoms for representing intermolecular electrostatic interactions related to halogen bonding and their substituent effects. *Journal of Computational Chemistry*, 31(1): 107–116, 2010. ISSN 1096-987X. doi: 10.1002/jcc.21302.

[14] H.J. Jahromi and K. Eskandari. Halogen bonding: a theoretical study based on atomic multipoles derived from quantum theory of atoms in molecules. *Structural Chemistry*, pages 1–7, 2012. ISSN 1040-0400. doi: 10.1007/s11224-012-0156-2.

[15] Peter Politzer, Jane S. Murray, and MonicaC. Concha. σ-hole bonding between like atoms; a fallacy of atomic charges. *Journal of Molecular Modeling*, 14:659–665, 2008. ISSN 1610-2940. doi: 10.1007/s00894-008-0280-5.

[16] Araz Jakalian, Bruce L. Bush, David B. Jack, and Christopher I. Bayly. Fast, efficient generation of high-quality atomic charges. am1-bcc model: I. method. *Journal of Computational Chemistry*, 21 (2):132–146, 2000. ISSN 1096-987X. doi: 10.1002/(SICI)1096-987X(20000130)21:2¡132::AID-JCC5¿3.0.CO;2-P.

[17] D.A. Case, T.A. Darden, T.E. III Cheatham, C.L. Simmerling, J. Wang, R.E. Duke, R. Luo, R.C. Walker, W. Zhang, K.M. Merz, B. Roberts, S. Hayik, A. Roitberg, G. Seabra, J. Swails, A.W. Goetz, I. Kolossvry, K.F. Wong, F. Paesani, J. Vanicek, R.M. Wolf, J. Liu, X. Wu, S.R. Brozell, T. Steinbrecher, H. Gohlke, Q. Cai, X. Ye, J. Wang, M.-J. Hsich, G. Cui, D.R. Roe, D.H. Mathews, M.G. Seetin, R. Salomon-Ferrer, C. Sagui, V. Babin, T. Luchko, S. Gusarov, A. Kovalenko, and P.A. Kollman. Amber 12, 2012.

[18] U. Chandra Singh and Peter A. Kollman. An approach to computing electrostatic charges for molecules. *Journal of Computational Chemistry*, 5 (2):129–145, 1984. ISSN 1096-987X. doi: 10.1002/jcc.540050204.

[19] Christopher I. Bayly, Piotr Cieplak, Wendy Cornell, and Peter A. Kollman. A well-behaved electrostatic potential based method using charge restraints for deriving atomic charges: the resp model. *The Journal of Physical Chemistry*, 97(40):10269–10280, 1993. doi: 10.1021/j100142a004.

[20] S.F. Boys and F. Bernardi. The calculation of small molecular interactions by the differences of separate total energies. some procedures with reduced errors. *Molecular Physics*, 19(4):553–566, 1970. doi: 10.1080/00268977000101561.

[21] Alex A. Granovsky. Firefly version 7.1.g.

[22] H.-J. Werner, P. J. Knowles, G. Knizia, F. R. Manby, M. Schütz, P. Celani, T. Korona, R. Lindh, A. Mitrushenkov, G. Rauhut, K. R. Shamasundar, T. B. Adler, R. D. Amos, A. Bernhardsson, A. Berning, D. L. Cooper, M. J. O. Deegan, A. J. Dobbyn, F. Eckert, E. Goll, C. Hampel, A. Hesselmann, G. Hetzer, T. Hrenar, G. Jansen, C. Köppl, Y. Liu, A. W. Lloyd, R. A. Mata, A. J. May, S. J. McNicholas, W. Meyer, M. E. Mura, A. Nicklass, D. P. O'Neill, P. Palmieri, D. Peng, K. Pflüger, R. Pitzer, M. Reiher, T. Shiozaki, H. Stoll, A. J. Stone, R. Tarroni, T. Thorsteinsson, and M. Wang. Molpro, version 2012.1, a package of ab initio programs, 2012. see http://www.molpro.net.

[23] Noel O'Boyle, Michael Banck, Craig James, Chris Morley, Tim Vandermeersch, and Geoffrey Hutchison. Open babel: An open chemical toolbox. *Journal of Cheminformatics*, 3(1):33, 2011. ISSN 1758-2946. doi: 10.1186/1758-2946-3-33.

[24] URL `http://openbabel.org`.

[25] Emma Sigfridsson and Ulf Ryde. Comparison of methods for deriving atomic charges from the electrostatic potential and moments. *Journal of Computational Chemistry*, 19(4):377–395, 1998. ISSN 1096-987X. doi: 10.1002/(SICI)1096-987X(199803)19:4¡377::AID-JCC1¿3.0.CO;2-P.

[26] E. Anderson, Z. Bai, C. Bischof, S. Blackford, J. Demmel, J. Dongarra, J. Du Croz, A. Greenbaum, S. Hammarling, A. McKenney, and D. Sorensen. *LAPACK Users' Guide*. Society for Industrial and Applied Mathematics, Philadelphia, PA, third edition, 1999. ISBN 0-89871-447-8 (paperback).

[27] Wendy D. Cornell, Piotr Cieplak, Christopher I. Bayly, Ian R. Gould, Kenneth M. Merz, David M. Ferguson, David C. Spellmeyer, Thomas Fox, James W. Caldwell, and Peter A. Kollman. A second generation force field for the simulation of proteins, nucleic acids, and organic molecules. *Journal of the American Chemical Society*, 117(19):5179–5197, 1995. doi: 10.1021/ja00124a002.

[28] Matthew Clark, Richard D. Cramer, and Nicole Van Opdenbosch. Validation of the general purpose tripos 5.2 force field. *Journal of Computational Chemistry*, 10(8):982–1012, 1989. ISSN 1096-987X. doi: 10.1002/jcc.540100804.

In: Models in Bioscience and Materials Research
Editor: Kh. T. Kholmurodov
ISBN: 978-1-62808-052-0

Chapter 5

THE STUDY OF PHASE DIAGRAM OF THE SURFACTANT/WATER SYSTEM BY MOLECULAR DYNAMICS SIMULATION

***Aram A. Shahinyan*[1,2,*], *L. H. Arsenyan*[1,2] *and A. H. Poghosyan*[1,2]**
[1]International Scientific-Educational Center
National Academy of Sciences , Yerevan
Republic of Armenia
[2]Institute of Applied Problems of Physics
NASRA

Abstract

Molecular dynamics study of complex molecular systems is an important addition to the physical experiment. In this paper we have performed 300 ns Molecular Dynamics (MD) to study the formation of spherical, cylindrical and flat micelles in the micellar and liquid-crystalline phases of amphiphile/water system. Some important parameters have been obtained which characterize the dynamic structure of micelles. It was found that diffusion rate of amphiphilic molecules in spherical, cylindrical and flat micelles are significantly different of each other because of the specific conditions of localization of molecules in these structural units.

Keywords: Molecular Dynamics Simulation, Spherical Micelle, Cylindrical Micelle, Plane Micelle, Lyotropic Liquid Crystal

*E-mail address:artsha@sci.am

1. Introduction

Researches of scientists of recent years have shown that methods and opportunities of information technologies can be quite effectively used to study the structure and properties of complex, multi-component, both biological and chemical systems in dynamics [1-7] and the specialized algorithms and calculation tools have created, one of the most common of which is the method of "Molecular Dynamics" - MD [8]. The main goal of the present paper is to investigate the mechanism of formation of the basic structural units formed in the amphiphilie/water system using MD simulation and the main objectives of this work is to study the dynamic structure of basic molecular formations: spherical, cylindrical and plane micelles throughout the phase diagram of this system. The investigations were performed with aqueous solution of commercially important product - sodium pentadecylsulphonate (SPDS) in water. At present, computational studies of these types of systems are very crucial, because they will be able to answer on many questions related to scientific and practical applications of amphiphilic compounds.

2. Construction and Simulation Data

For the system construction we have created a molecule of SPDS using Hyperchem (Hypercube Inc.) software and MDesigner [9, 14] by random replication we have received system consisting 128 molecules with 128 sodium counterions. The system was inserted into water bulk with various amounts of water molecules of TIP3P [10] model. For the system, the constant temperature and pressure were set correspondingly to 323 K and normal 1atm. The Langevin dynamics [11] with 5ps^{-1} damping coefficient was used for constant temperature control and the constant pressure was controlled by using the Langevin piston Nose-Hoover method [12]. For non-bonded full electrostatic interactions between atom pairs cutoff the parameter was set to 14 Å and PME with the precision of 10^{-6} was used. The coordinates and velocities were saved each 0.1 ns and the visual representation performed by using VMD package [13]. System energy minimization and simulation run (NPT ensemble) were done with 1 fs time step, using GROMACS software code with force field on parallel Linux cluster. The force field parameters of SPDS molecule were generated using the Dundee PRODRG server [14]. The investigations were carried out using the GROMACS software package. The duration of the computational experiment was 300 ns.

3. Results and Discussion

There are various theories about the mechanism of the formation of micelles in solutions of amphiphilic compounds. In some of them the process of micelle formation [15,16] is considered as a phase transition, where at a certain concentration of amphifile - a critical micelle concentration (CMC), micelles appear abruptly. In accordance with another hypothesis, formation of micelles is one stage process [17], when the micellar aggregates and individual surfactant molecules in the solution exist simultaneously and are in equilibrium. On the other hand it is known that the formation of micelles is mainly due to hydrophobic interactions [18], which play an important role in the chemistry and biology.

Let us try to follow the formation of spherical, cylindrical and plane micelles in aqueous solution of Sodium pentadecylsulphonate (SPDS). Figure 1 shows the formation of spherical micelles in aqueous solution SPDS with aconcentration 263 mM. As we can see at this concentration of SPDS spherical micelles are formed.

As the figure shows formation of micelles is the stepwise process with the formation of intermediate pre-micellar associates. As is known [18], the forces leading to the formation of micelles have a hydrophobic nature. At the same time, because of the presence of water soluble and water insoluble sites in amphiphilic molecules, the polar groups of SPDS molecules are localized at the surface and hydrocarbon tails inside the micelle. At higher concentrations of amphiphilic substances a transition from the micellar to the liquid-crystalline phase is observed. At the same time, the spherical micelles are deformed in cylindrical and flat micelles. The formation of cylindrical micelle investigated at the concentration 526 mmol/l and the flat micelles at the concentration of 3 mol/l. Fig.2 shows the formation of cylindrical from spherical and flat from cylindrical micelles obtained by the MDsimulation.

Let us try to describe different phases of SPDS / water system through the mobility of SPDS molecules within the corresponding types of micelles. For studying the motion of SPDS molecules in micelles thc diffusion of molecules within the micelles was studied by the method of molecular simulation. The diffusion of the molecules was studied by measuring the time dependence of the standard deviation of the polar groups of SPDS molecules (MSD) on the

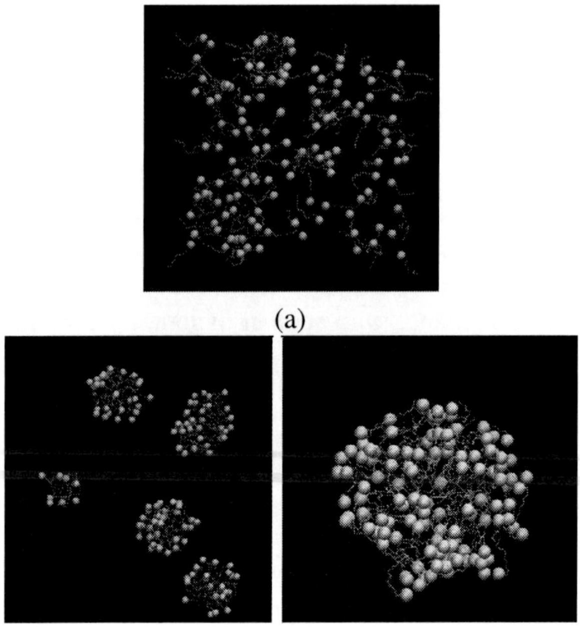

(a)

(b) (c)

Figure 1. Formation of spherical micelles: a - start of micellization from SPDS molecular solution in water, b - after 4.5 ns, c - after 46 ns (the end of micelle formation). Water molecules are not shown.

surfaces of micelles:

$$MSD = \sum_{all} < |\vec{r}(t) - \vec{r}(0)|^2 >, \quad (1)$$

were $\vec{r}(0)$ and $\vec{r}(t)$ are radius vectors defining the location of polar groups of SPDS molecules on the surface of micelles at the beginning and after the simulation time t.

There can be two basic types of molecular diffusion in a micelles depending on their form: the volume diffusion (in the case of spherical and cylindrical micelles) and lateral diffusion in the flat micelles. In all three forms of micelles, the polar groups of amphiphilic molecules are moving on the surface of micelles.

The coefficient of diffusion is defined as:

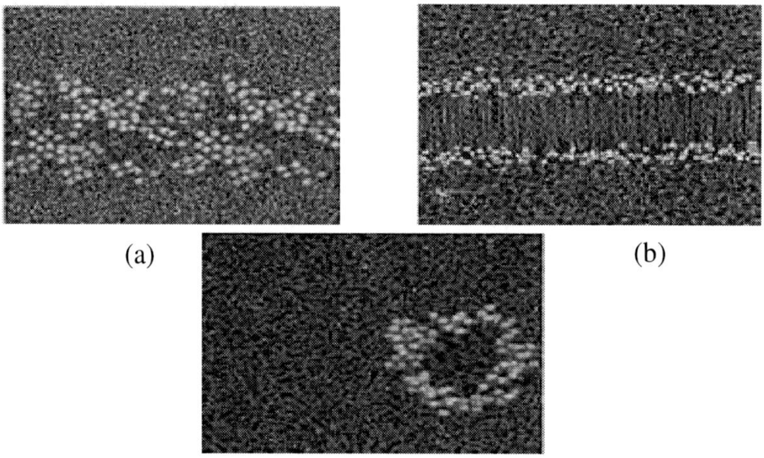

(a) (b) (c)

Figure 2. Formation of cylindrical (a) and flat (b) micelles, c-plane cross section of cylindrical micelle.

for spherical or cylindrical micelles

$$D = MSD/6t, \tag{2a}$$

for flat micelles

$$D = MSD/4t, \tag{2b}$$

Fig. 3 shows the dependence of MSD values of SPDS molecules in spherical, cylindrical and flat micelles. By using linear fit (Fig.3) and formula (2) diffusion coefficients of polar groups of SPDS molecules at the surfaces of micelles were calculated (Tab.1).

As seen from Tab.1, polar groups of SPDS molecules, hence the molecule as a whole,most rapidly are moving in the flat micelle and most slowlyin the spherical ones.

Perhaps the reason for this is that (see Fig.4), unlike the flat micelles, in spherical and cylindrical micelles terminal hydrocarbon groups of SPDS molecules are located in the narrow space at the center of sphere or cylinder, so that there is inhibition of their movement, there by reducing the speed of whole molecules.

Thus by the method of molecular simulation, observed presence of correlation between the phase state of amphiphilic/ water system and of the regularity of amphiphilic molecules diffusion.

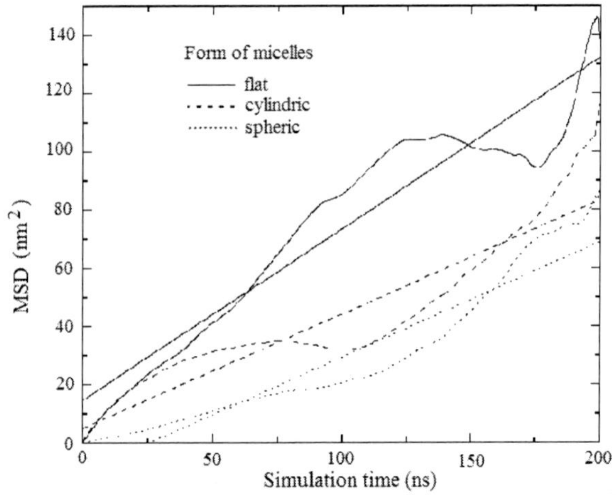

Figure 3. The dependence of MSD of polar head groups of SPDS molecules on the time of simulation at the surfaces of spherical, cylindrical and flat micelles. Each line is alsoapproximated by straight line.

Table 1. Coefficients of diffusion (cm^2/s) for polar head groups of SPDS molecules on the surfaces of spherical, cylindrical and flat micelles

Form of micelles	Diffusion coefficient
Spherical micelle (Micellar phase)	0.518×10^{-6} cm^2/s
Cylindrical micelle (Liquid Crystalline phase)	0.898×10^{-6} cm^2/s
Flat micelle (Liquid Crystalline phase)	0.944×10^{-6} cm^2/s

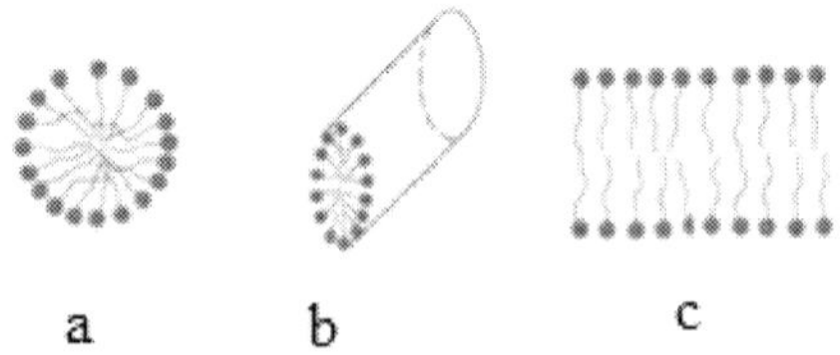

Figure 4. a-spherical, b-cylindrical and c-flat micelles.

References

[1] Poghosyan A.H., Gharabekyan H.H., Shahinyan A.A., "Molecular Dynamics Simulation of DMPC/DPPC Mixed Bilayers". *IJMPC*, 73, (2007), pp. 73-89.

[2] Mori K., Hata M., Neya S., Hoshino T. "MD simulation of asymmetric phospholipid biayers with ions and cholesterols". *Chem-Bio Informatics J.*, 4, (2004), pp.15-26.

[3] Robinson A. J., Richards W. G., Thomas P. J., Hann M. M. "Behvior of chlesterol and it's effect on head group and chain conformations in lipid bilayers: a molecular dynamics study". *Biophys. J.*, 68, (1995), pp.164-170.

[4] Berkowitz M. L. "Detailed Molecular Dynamics Simulations of Model Biological Membranes Containing Cholesterol". *Bioph. Bioch. Acta,* 1788, (2009), pp.86-96.

[5] Soumana O.S., Garnier N. "Molecular Dynamics simulation for the prediction of transmembrane helix-helix heterodimers assembly". *Eur. Biophys. J.*, 36, (2007), pp.1071-1082.

[6] Zhang J., Lazaridis T. "Transmembrane Helix Association Affinity Can Be Modulated by Flanking and Noninterfacial Residues",*Biophys. J.,* 96, (2009), pp.4418-4427.

[7] Eilers M., Patel A.B., Liu W., Smith S.O. "Comparison of Helix Interactions in Membrane and Soluble -Bundle Protein". *Biophys. J.*, 82, (2002), pp.2720-2736.

[8] Smith S.O., Song D., Shekar S., Groesbeek M., Ziliox M., Aimoto S. "Structure of the transmembrane dimer interface of glycophorin A in membrane bilayers.Biochemistry". *Biochemistry*, 40, (2001), pp.6553-6558.

[9] Schuettelkopf, A. W.; van Aalten, D. M. F. "PRODRG a tool for high-throughput crystallography of protein-ligand complexes". *Acta Crystallogr. D.*, 60, (2004), pp.1355-1363.

[10] Jorgensen W.L., Chandrasekhar J., Medura J.D., Imprey R.W. and Klein M.L. "Comparison of Simple Potential Functions for Simulating Liquid Water", *J. Chem. Phys.*, 79, (1983), pp.928-935.

[11] Levy R.M., McCommon J.A. and Karplus M. "Diffusive Langevin dynamics of model alkanes". *Chem. Phys. Lett.,* 64, (1979), pp.1-4.

[12] Feller S. E., Zheng Y. H., Pastor R.W. and Brooks B.R. "Constant pressure molecular dynamics simulation-the Langevin piston method", *J. Comp. Phys.,* 103, (1995), pp.4613-4621.

[13] Humprey W., Dalke A. and Schulten K. "VMD: visual molecular dynamics". *J. Molec. Graphics,* 14, (1996), pp.33-38.

[14] Shahinyan A.A., Poghosyan A.H., Eghiazaryan G.A., Garabekyan H.H. "Mdesigner: A Software Tool, for Construction of Biomembrans", *Electronic J. Nat.Sci,* 1, (2004), pp.56-60.

[15] Minasyants M. K., Zakharyan V .A.,Shahinyan A. A. , Chistiakov I. G. "X-ray analysis of Na-dodecylsulfatewater system", *Crystallografia,* 24, (1979), pp.319-321 (in Russian).

[16] Corkill J. M., Goodman J. F., Walker T., Wyer J., "The Multiple Equilibrium Model of Micelle Formation". *Proceedings of the Royal Society of London. Series A, Mathematical and Physical Sciences,* 312, (1969), Issue 1509, pp.243-255.

[17] Mukerjee P., "The nature of the association equilibria and hydrophobic bonding in aqueous solution of association colloids". *Adv. Colloid Iterface Sci.,* 1, (1967), pp.241-275.

[18] C.Tenford, "Hydrophobic Effect", 2nd ed., John Wiley, New York (1980).

In: Models in Bioscience and Materials Research ISBN: 978-1-62808-052-0
Editor: Kh. T. Kholmurodov

Chapter 6

ALL-ATOM MULTIBARIC-MULTITHERMAL MOLECULAR DYNAMICS SIMULATIONS OF PROTEINS

Hisashi Okumura[1,2,*]
[1]Research Center for Computational Science
Institute for Molecular Science, Okazaki, Japan
[2]Department of Structural Molecular Science
The Graduate University for Advanced Studies, Okazaki, Japan

Abstract

A review of the multibaric-multithermal molecular dynamics algorithm is presented. In this algorithm, two-dimensional random walks not only in the potential-energy space but also in the volume space are realized. This algorithm is a powerful tool to discuss the temperature dependence and pressure dependence of biomolecules. Applications to an alanine dipeptide and chignolin in explicit water solvent are presented. The partial molar enthalpy change and partial molar volume change of unfolding can be calculated. Illustrating typical local-minimum free-energy conformations, folding and unfolding pathways can be revealed. We can also discuss the mechanism of temperature and pressure denaturation.

Keywords: Molecular Dynamics Simulation, Multibaric-Multithermal Ensemble, Alanine Dipeptide, Chignolin

*E-mail address: hokumura@ims.ac.jp

1. Introduction

Molecular dynamics (MD) simulation is a standard tool to investigate the conformational change of biomolecules such as proteins at atomic level. However, biomolecules have complicated free energy surfaces with many local minima. Thus, conventional MD simulations in physical ensembles, such as the canonical [1, 2, 3] and isobaric-isothermal [4] ensemble, tend to get trapped in these local-minimum states. One of the powerful techniques to avoid this difficulty is generalized-ensemble algorithms [5, 6, 7, 8, 9] such as the multicanonical algorithm [10, 11, 12, 13]. In the multicanonical ensemble, a free one-dimensional random walk is realized in the potential-energy space and a simulation does not get trapped in free-energy-minimum states.

However, because the multicanonical simulation is performed in a fixed volume, neither the pressure dependence nor temperature dependence at certain pressure can be investigated as in experiments. To overcome this difficulty, the author have proposed the multibaric-multithermal algorithm [14, 15, 16, 17, 18]. In this algorithm, two-dimensional random walks are realized both in the potential-energy space and in the volume space, so that the temperature and pressure dependence can be discussed.

In this review article, the multibaric-multithermal algorithm is explained and applications of this algorithm to an alanine dipeptide and chignolin in explicit water are presented. The multicanonical and multibaric-multithermal algorithms are described in Sec. 2. Applications to an alanine dipeptide and chignolin are presented in Sec. 3. Section 4 is devoted to conclusions.

2. Methods

2.1. Multicanonical Algorithm

In the canonical ensemble, the distribution $\mathcal{P}_{NVT}(E)$ of total potential energy E is given by

$$\mathcal{P}_{NVT}(E) = n(E)\mathrm{e}^{-\beta_0 E} , \tag{1}$$

where $n(E)$ is the density of states, β_0 is the inverse of the product of the Boltzmann constant k_{B} and absolute temperature T_0, at which simulations are performed. This ensemble has a bell-shaped distribution in the potential-energy space.

In the multicanonical ensemble [10, 11, 12, 13], every state is sampled with a weight factor $W_{\rm muca}(E) \equiv \exp\{-\beta_0 E_{\rm muca}(E)\}$ so that a uniform distribution of E may be obtained:

$$\mathcal{P}_{\rm muca}(E) = n(E)\, W_{\rm muca}(E) = \text{constant}\ , \tag{2}$$

where $E_{\rm muca}(E)$ and $W_{\rm muca}(E)$ are referred to as the multicanonical energy and the multicanonical weight factor, respectively. The difference between $E_{\rm muca}$ and E is written as $\delta E(E)$: $E_{\rm muca}(E) \equiv E + \delta E(E)$. The difference $\delta E(E)$ is characteristic of the multicanonical simulation. The case of $\delta E(E) = 0$ gives the regular canonical ensemble. The multicanonical simulation covers a wide range of E and escapes from local-minimum free-energy states. The expectation value of a physical quantity A at temperature T is obtained by reweighting techniques [19] from

$$\langle A \rangle_{NVT} = \frac{\langle A(E) \exp[\beta_0\{E + \delta E(E)\}] \exp\{-\beta E\}\rangle_{\rm muca}}{\langle \exp[\beta_0\{E + \delta E(E)\}] \exp\{-\beta E\}\rangle_{\rm muca}}\ , \tag{3}$$

where $\langle \cdots \rangle_{\rm muca}$ is the multicanonical ensemble average. Because of the random walks in the potential-energy space, we can calculate physical quantities in a wide range of T. The multicanonical weight factor $W_{\rm muca}(E)$ is not *a priori* known and has to be determined, for example, by the iterations of short simulations [20, 21, 22] and the Wang-Landau techniques [23].

2.2. Multibaric-Multithermal Algorithm

In the isobaric-isothermal ensemble [4], the distribution $\mathcal{P}_{NPT}(E, V)$ of potential energy E and volume V is given by

$$\mathcal{P}_{NPT}(E, V) = n(E, V) e^{-\beta_0 H}\ , \tag{4}$$

where $n(E, V)$ is the density of states as a function of E and V and H is the "enthalpy" (without the kinetic energy contributions): $H = E + P_0 V$. Here, P_0 is the pressure at which simulations are performed. This ensemble has bell-shaped distributions both in the potential-energy space and in the volume space, as shown in Fig. 1(a). In order to obtain the isobaric-isothermal ensemble, the combination of the Nosé thermostat [1, 2] and the Andersen barostat [4] is frequently employed.

In the multibaric-multithermal ensemble [14, 15, 16, 17, 18], every state is sampled with a weight factor $W_{\rm mbt}(E,V) \equiv \exp\{-\beta_0 H_{\rm mbt}(E,V)\}$ so that a uniform distribution of both E and V, as shown in Fig. 1(b), may be obtained:

$$\mathcal{P}_{\rm mbt}(E,V) = n(E,V)W_{\rm mbt}(E,V) = {\rm constant}\ . \tag{5}$$

Here, $W_{\rm mbt}(E,V)$ and $H_{\rm mbt}$ are referred to as the multibaric-multithermal weight factor and the multibaric-multithermal enthalpy, respectively. The difference between $H_{\rm mbt}$ and H is written as $\delta H(E,V)$: $H_{\rm mbt}(E,V) = H + \delta H(E,V)$. The difference $\delta H(E,V)$ is therefore characteristic of the multibaric-multithermal simulation. The case of $\delta H(E,V) = 0$ gives the regular isobaric-isothermal ensemble.

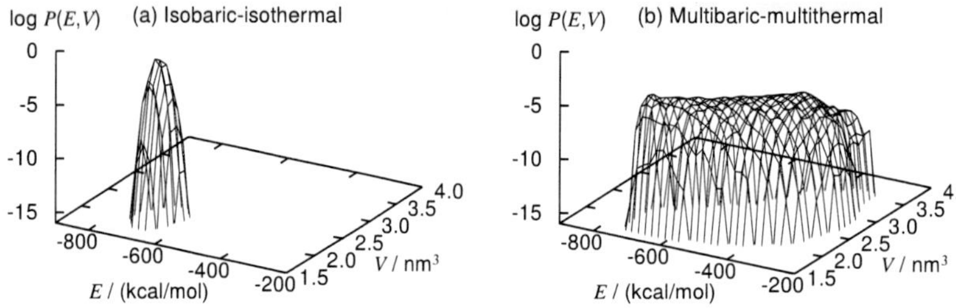

Figure 1. Distributions $\mathcal{P}(E,V)$ of potential energy E and volume V of an alanine dipeptide in explicit water (a) by the isobaric-isothermal MD simulation and (b) by the multibaric-multithermal MD simulation.

The equations of motion in the multibaric-multithermal ensemble based on the Nosé thermostat [1, 2] and the Andersen barostat [4] are given by

$$\dot{\mathbf{r}}_i = \frac{\mathbf{p}_i}{m_i} + \frac{\dot{V}}{3V}\mathbf{r}_i\ , \tag{6}$$

$$\dot{\mathbf{p}}_i = \left(1 + \frac{\partial \delta H}{\partial E}\right)\mathbf{F}_i - \left(\frac{\dot{V}}{3V} + \frac{\dot{s}}{s}\right)\mathbf{p}_i\ , \tag{7}$$

$$\dot{V} = s\frac{P_V}{W}\ , \tag{8}$$

$$\dot{P_V} = s\left[\frac{1}{3V}\left\{\sum_{i=1}^{N}\frac{\mathbf{p}_i^2}{m_i} + \left(1 + \frac{\partial \delta H}{\partial E}\right)\sum_{i=1}^{N}\mathbf{F}_i\cdot\mathbf{r}_i\right\} - \left(P_0 + \frac{\partial \delta H}{\partial V}\right)\right] \tag{9}$$

$$\dot{s} = s\frac{P_s}{Q} , \tag{10}$$

$$\dot{P}_s = \sum_{i=1}^{N+M} \frac{\mathbf{p}_i^2}{m_i} - gk_{\mathrm{B}}T_0 , \tag{11}$$

where $\mathbf{r}_i$ is the coordinate, $\mathbf{p}_i$ is the momentum, s is the additional degree of freedom for the Nosé thermostat, the dot ($\dot{}$) stands for the real time derivative d/dt, and $\mathbf{F}_i$ stands for the force acting on atom i. The variables P_V and P_s are the conjugate momenta for V and s, respectively. The constant m_i is the mass of atom i. The constants W and Q are the artificial "mass" related to V and s, respectively. Performing the MD simulation by the equations of motion, the multibaric-multithermal distribution $\mathcal{P}_{\mathrm{mbt}}(E, V)$ in Eq. (5) is obtained.

In the case of $\delta H(E,V) = \partial\delta H(E,V)/\partial E = \partial\delta H(E,V)/\partial V = 0$, the equations of motion in Eqs. (6)–(11) become those for the regular isobaric-isothermal MD simulation of the Nosé-Andersen formulation.

After an optimal weight factor $W_{\mathrm{mbt}}(E, V)$ is determined, for example, by the iterations of short simulations [20, 21, 22] or by the Wang-Landau techniques [23], a long production run is performed for data collection. The reweighting techniques [19] are used for the results of the production run to calculate the isobaric-isothermal-ensemble averages. The expectation value of a physical quantity A at T and P is then obtained by

$$\langle A \rangle_{NPT} = \frac{\langle A(E,V) W_{\mathrm{mbt}}^{-1}(E,V) \mathrm{e}^{-\beta(E+PV)} \rangle_{\mathrm{mbt}}}{\langle W_{\mathrm{mbt}}^{-1}(E,V) \mathrm{e}^{-\beta(E+PV)} \rangle_{\mathrm{mbt}}} , \tag{12}$$

where $\langle \cdots \rangle_{\mathrm{mbt}}$ is the multibaric-multithermal ensemble average. Because of the random walks both in the potential-energy space and in the volume space, we can calculate physical quantities in wide ranges of T and P.

3. Examples of Simulation Results

3.1. Alanine Dipeptide

Now two examples of the simulation results by the multibaric-multithermal MD algorithm are presented. First example is an alanine dipeptide in explicit water solvent. Alanine dipeptide has been studied theoretically extensively as a model biomolecule because of its simplicity. The system consists of one alanine dipeptide molecule ((*S*)-2-(acetylamino)-*N*-methylpropanamide) and 63 water molecules [24, 25]. This initial conformation is shown in Fig. 2.

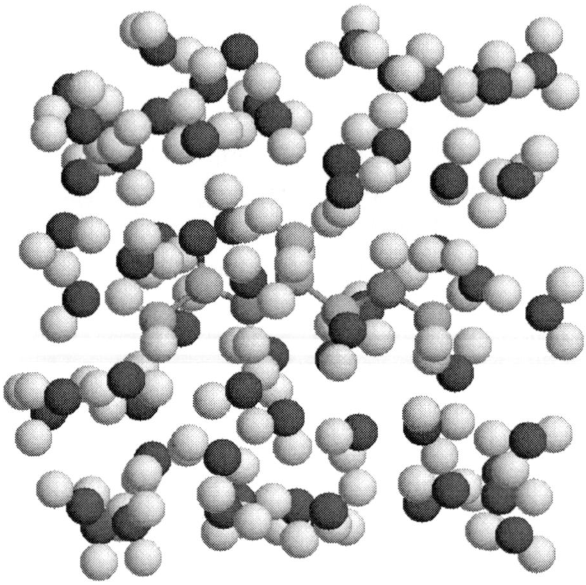

Figure 2. The initial conformation of alanine dipeptide for the MD simulations. The dihedral angles are $\phi = \psi = 180°$. The figure was created with RasMol [26].

The AMBER parm96 force field [27] was used for the alanine dipeptide molecule and the TIP3P rigid-body [28] model was employed for the water molecules. The electrostatic potential was calculated by the Ewald method. The time step was taken to be Δt = 0.5 fs. The temperature was set as $T_0 = 600$ K and the pressure was set as $P_0 = 300$ MPa. The MD simulations were performed with the symplectic combination [29, 32] of the Nosé-Poincaré thermostat, the Andersen barostat [4], and the symplectic quaternion scheme. In order to obtain an optimal multibaric-multithermal weight factor $W_{\rm mbt}(E, V)$, two-dimensional versions of both the iterative method [20, 21, 22] and the Wang-Landau techniques [23], for multicanonical weight-factor determination were employed. An optimal weight factor was obtained after these preliminary simulations in total for 2.61 ns. Two sets of the initial values of the alanine-dipeptide backbone dihedral angles were prepared for the data-production runs. Initial condition 1 was $\phi = \psi = 180°$ and initial condition 2 was $\phi = 180°$ and $\psi = 0°$. We then performed a long multibaric-multithermal MD simulation for data collection. The simulation time was 20 ns for each initial condition. The isobaric-isothermal MD simulations of the same system were also performed for comparisons.

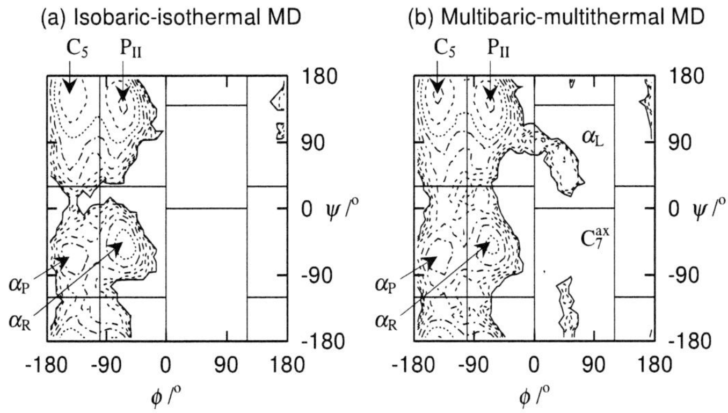

Figure 3. Contour maps of the logarithms of the probability distributions $\mathcal{P}(\phi, \psi)$ of the backbone dihedral angles ϕ and ψ at T = 240 K and P = 0.1 MPa. (a) $\mathcal{P}(\phi, \psi)$ obtained from initial condition 1 by the conventional isobaric-isothermal MD simulation. (b) $\mathcal{P}(\phi, \psi)$ obtained from initial condition 1 by the reweighting techniques from the results of the multibaric-multithermal MD simulation.

Figure 3 shows the contour maps of the distributions $\mathcal{P}(\phi, \psi)$ at T = 240 K and P = 0.1 MPa by the MD simulations started from the initial condition 1. The distribution $\mathcal{P}(\phi, \psi)$ by the multibaric-multithermal MD were obtained using the reweighting techniques. The isobaric-isothermal distribution $\mathcal{P}(\phi, \psi)$ has only four peaks of the α_R, α_P, P_{II}, and C_5 states, as shown in Figs. 3(a). On the other hand, the multibaric-multithermal MD simulation provides correct distributions $\mathcal{P}(\phi, \psi)$ with all the six peaks of the α_R, α_P, P_{II}, C_5, α_L, and C_7^{ax} states, as shown in Figs. 3(b). These results mean that the multibaric-multithermal simulation can get over the free energy barriers and has a high sampling efficiency.

Table 1. Differences in partial molar enthalpy ΔH / (kJ/mol) and in partial molar volume ΔV / (cm^3 mol^{-1}) of the C_5 and α_R states from those of the P_{II} state calculated by the multibaric-multithermal (MUBATH) MD simulations. Raman experimental data are taken from Ref. [33]

	ΔH		ΔV	
State	MUBATH MD	Raman exp.	MUBATH MD	Raman exp.
C_5	1.8 ± 0.5	2.5 ± 0.3	-0.3 ± 0.7	0.1 ± 0.3
α_R	4.4 ± 1.1	4.4 ± 1.5	-2.3 ± 1.3	1.1 ± 0.2

The volume under the surface $\mathcal{P}(\phi, \psi)$ around each peak corresponds to the population f of each state. The ratio of each state population f against the P_{II} state population was calculated as a function of $1/T$ at $P = 0.1$ MPa and as a function of P at $T = 298$ K. The difference in partial molar enthalpy ΔH and the difference in partial molar volume ΔV can be obtained by

$$\begin{aligned} \Delta H &= -R \left[\frac{\partial \log(f_{C_5}/f_{P_{II}})}{\partial(1/T)} \right]_P , \\ \Delta V &= -RT \left[\frac{\partial \log(f_{C_5}/f_{P_{II}})}{\partial P} \right]_T , \end{aligned} \tag{13}$$

where R is the gas constant: $R = 8.3145\ \mathrm{J/(mol \cdot K)}$. The values of ΔH and ΔV are listed in Table 1. Partial molar enthalpy and partial molar volume are important quantities in solution chemistry, which describes the change of the ratio of each state population due to temperature and pressure changes. The differences ΔH and ΔV of the C_5 state and ΔH of the α_R state are consistent in the multibaric-multithermal MD and the Raman spectroscopy experimental data [33]. The value of ΔV of the α_R state, however, is smaller than that of the Raman spectroscopy.

The multicanonical simulation can escape from local-minimum free-energy states, but cannot change volume. The conventional isobaric-isothermal MD simulation can change volume, but is trapped in a local free energy minimum. Therefore, ΔH and ΔV cannot be calculated by these simulation algorithms. The multibaric-multithermal algorithm has the advantages of both algorithms. This is the first simulational work to calculate ΔH and ΔV of biomolecules.

3.2. Chignolin

The next application is a chignolin in explicit water solvent [34]. Chignolin is a 10 residue protein (GYDPETGTWG) of which the native state is a β-hairpin structure [35]. The N and C termini of the protein were left uncapped as in the experiment by Honda et al. [35]. That is, the N terminus and C terminus have a positive and negative electric charge, respectively. The system is consisting of one chignolin molecule, 902 water molecules, and two sodium ions Na^+ as counter ions. The initial values of the chignolin backbone dihedral angles were $\phi = \psi = 180°$ except for proline. The initial dihedral-angle values of PRO4 were set to $\phi = -60°$ and $\psi = 180°$. AMBER parm99SB force field [37]

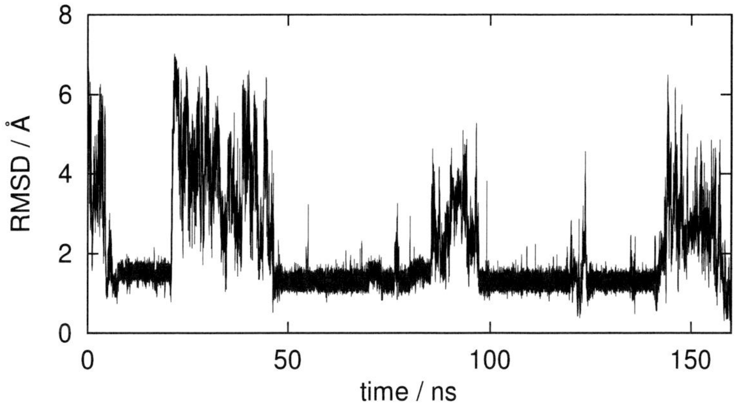

Figure 4. The time series of root mean square deviation (RMSD). Typical snapshots are also illustrated, which were created with RasMol [26].

was used for the chignolin molecule and the TIP3P [28] rigid-body model was used for the water molecules. A cubic unit cell was employed with periodic boundary conditions. The electrostatic potential was calculated by the Ewald method. Cutoff distance was $r_c = 12$ Å for both electrostatic and Lennard-Jone potential. The combination [32] of the Nosé-Hoover thermostat [1, 2, 3], the Andersen barostat [4], and the symplectic quaternion scheme [29, 30] was used for the rigid-body water molecules. Reversible multiple time scale molecular dynamics techniques [31] were also applied. The time step was taken to be Δt

= 0.5 fs for the protein atoms and Δt = 4.0 fs for the water molecules. Because the symplectic rigid-body algorithm was used for the water molecules here, Δt was able to be taken as long as 4.0 fs [32].

Figure 4(a) shows time series of root mean square deviation (RMSD). The RMSD was calculated for the backbone N, C_α, and C atoms with respect to the reference conformation. The model 1 of the nuclear magnetic resonance (NMR) structure (PDB ID code 1UAO) was employed for the reference conformation. Because the initial condition is an extended structure, the initial value of the RMSD was as large as RMSD=6.39 Å. The RMSD became small after 6 ns and became large again after 21 ns. The RMSD decreased and increased repeatedly in this way. Decrease and increase of RMSD mean that folding and unfolding events occurred, as typical snapshots in Fig. 4(a). The folding events were observed five times and the unfolding events occurred four times during this simulation.

Temperature and pressure dependences of the fraction of the folded chignolin f_{fold} are shown in Fig. 5. The fraction of the folded protein f_{fold} decreases as temperature and /or pressure increases. The partial molar enthalpy change ΔH and the partial molar volume change ΔV on unfolding is calculated by

$$\Delta H = R\left[\frac{\partial \log\{f_{\text{fold}}/(1-f_{\text{fold}})\}}{\partial(1/T)}\right]_P, \tag{14}$$

$$\Delta V = RT\left[\frac{\partial \log\{f_{\text{fold}}/(1-f_{\text{fold}})\}}{\partial P}\right]_T. \tag{15}$$

Interpolating or extrapolating the temperature dependence of f_{fold}, the derivatives in Eqs. (14) and (15) were calculated. The partial molar enthalpy change ΔH was determined that $\Delta H = 24.1 \pm 4.9$ kJ/mol at $P = 0.1$ MPa, as listed in Table 2. Honda et al. determined ΔH by the CD spectroscopy and NMR experiments and their data distribute between 25.9 kJ/mol and 32.2 kJ/mol. The partial molar enthalpy change ΔH by the present multibaric-multithermal MD simulation is slightly lower than the experimental data, but shows a reasonably good agreement.

The partial molar volume change ΔV was determined that $\Delta V = -5.6 \pm 1.5$ cm^3/mol at $T = 298$ K, as listed in Table 2. Imamura and Kato obtained $\Delta V = -8.8$ cm^3/mol by their Fourier transform infraredspectrometer (FT-IR) experiments [36]. The absolute value of the present multibaric-multithermal MD result is slightly lower than their experimental data, but still agrees well.

The lower absolute values of ΔH and ΔV mean that chignolin does not unfold with the increasing temperature or pressure in the MD simulation as fast as in the experiment. A possible reason is that the Coulomb potential parameter or the electrostatic charge in the classical force field is too large. Reducing the absolute value of the electrostatic charge may evaluate the hydrogen bonds weaker so that the simulational temperature/pressure dependence of $f_{\rm fold}$ and ΔG may agree better with the experiment.

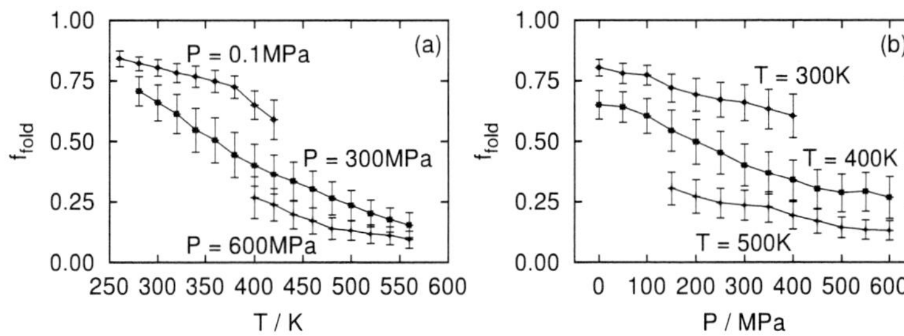

Figure 5. (a) Temperature dependence of fraction of the folded protein $f_{\rm fold}$ at $P = 0.1$, 300, and 600 MPa and (b) pressure dependence of $f_{\rm fold}$ at $T = 300$, 400, and 500 K obtained by the reweighting techniques from the results of the multibaric-multithermal MD simulation.

Table 2. Difference in the Gibbs free energy ΔG between the unfolded state and the folded state, difference in the partial molar enthalpy ΔH at $P = 0.1$ MPa, and difference in the partial molar volume ΔV at $T = 298$ K calculated by the multibaric-multithermal (MUBATH) MD simulation. Experimental data are taken from Refs. [35, 36]

Method	ΔG/(kJ/mol)	ΔH/(kJ/mol)	ΔV/(cm^3/mol)
MUBATH MD	3.5 ± 0.5	24.1 ± 4.9	-5.6 ± 1.5
Experiment	1.07 – 1.87	25.9 – 32.2	-8.8

The present multibaric-multithermal MD simulation sampled not only the β-hairpin structure but also 3_{10}-helix structure. Fig. 6 shows mean force potential as a function of $r(\mathrm{ASP3O} - \mathrm{GLY7N})$ and $r(\mathrm{GLU5O} - \mathrm{THR8N})$, which are the distance between ASP3O and GLY7N atoms and that between GLU5O and THR8N atoms, respectively. ASP3O and GLY7N atoms make a hydrogen bond at the β-hairpin structure and GLU5O and THR8N atoms make a hydrogen bond at the 3_{10}-helix structure. The potential of mean force has a global minimum state at $r(\text{ASP3O} - \text{GLY7N}) = 3.0$ Å and $r(\text{GLU5O} - \text{THR8N}) = 7.8$ Å at $T = 300$ K and $P = 0.1$ MPa, as shown in Fig. 6(a). This state is the native state, at which the chignolin folds into the β-hairpin structure. There also exists other states, which correspond to unfolded states. We can see blue and green distribution in a wide area at $T = 500$ K and $P = 600$ MPa, which means that the unfolded states increase, as shown in Fig. 6(b). There are several local-minimum free-energy states besides the global-minimum state. The state E is not a local-minimum states but a transition state between the states B and F or the states D and C.

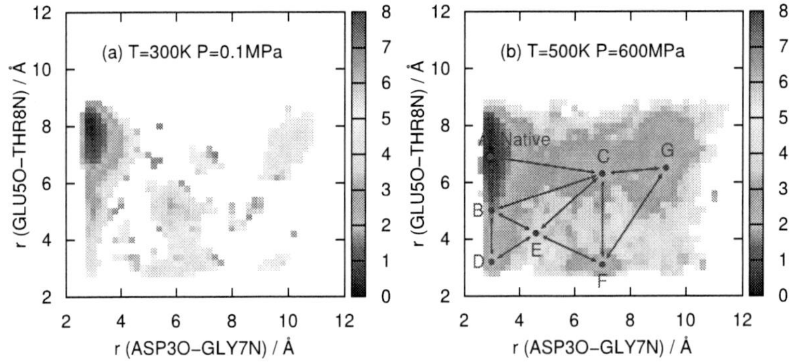

Figure 6. Potential of mean force (PMF) as a function of $r(\text{ASP3O} - \text{GLY7N})$ and $r(\text{GLU5O} - \text{THR8N})$ (a) at $T = 300$ K and $P = 0.1$ MPa and (b) at $T = 500$ K and $P = 600$ MPa obtained by the reweighting techniques from the results of the multibaric-multithermal MD simulation.

Typical conformations obtained by the multibaric-multithermal MD simulation at these states are illustrated in Fig. 7. At the global-minimum state A, chignolin forms the native β-hairpin structure. The local-minimum state B is obtained from the state A by bending the C terminus of the β-hairpin structure. The local-minimum state C is obtained from either state A or state B by bending both N and C terminus of the β-hairpin structure. This conformation looks like a Greek letter "Ω". If the N terminus of the state B does not bent like the state C, but make a turn, an α-helix structure is obtained at the state D. The transition sate E is obtained by breaking some of the hydrogen bonds of the α-helix structure at the state D. It can be obtained also from the state B or C. The local-minimum state F, a 3_{10}-helix structure, is obtained from the state E by forming hydrogen bonds in a different way from the α-helix structure at the state D. This 3_{10}-helix structure can also be attained from the Ω-like structure at the state C by making turn structures. Finally an extended structure is obtained at the state G if all hydrogen bonds are broken from the state C or F. Following these pathways, the native β-hairpin structure at the state A unfolds to the extended structure at the state G. When chignolin folds, it follows the reverse process to the native structure.

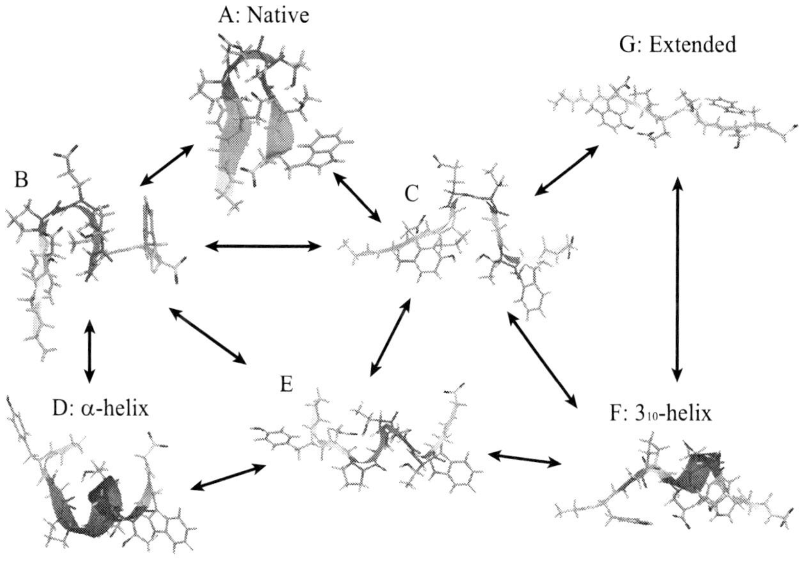

Figure 7. Folding and unfolding pathways of chignolin from the state A to G. These conformations are obtained by the multibaric-multithermal MD simulation. The N terminus and the C terminus are on the left-hand side and on the right-hand side, respectively. The figures were created with RasMol [26].

Conclusion

The multibaric-multithermal ensemble MD algorithm is reviewed. The multibaric-multithermal simulation performs a two-dimensional random walk both in the potential-energy space and in the volume space so that one can obtain various isobaric-isothermal ensemble averages at different temperatures and pressures from only one simulation run.

As applications of this algorithm, two examples of an alanine dipeptide and chignolin were presented. From the temperature and pressure dependences of each state, the differences in the partial molar enthalpy ΔH and the partial molar volume ΔV were calculated. The values of ΔH and ΔV calculated by molecular dynamics simulations showed reasonable agreement with experimental data. Potential of mean force of chignolin was also calculated. Free energy at the native state is low and those at the unfolded states are high at room temperature and pressure. Free energy at the unfolded states gets lower as temperature and/or pressure increases, reflecting that the β-hairpin structure unfolds at high temperature and/or pressure.

The multibaric-multithermal algorithm will thus be a powerful simulation technique to study the temperature and pressure dependences of biomolecules like proteins.

References

[1] Nosé S. "A molecular-dynamics method for simulations in the canonical ensemble". *Mol Phys* 52 (1984) 255-268.

[2] Nosé S. "A unified formulation of the constant temperature molecular-dynamics methods". *J Chem Phys* 81 (1984) 511-519.

[3] Hoover W.G. "Canonical dynamics: Equilibrium phase-space distributions". *Phys Rev A* 31 (1985) 1695-1697.

[4] Andersen H.C. "Molecular dynamics simulations at constant pressure and/or temperature". *J Chem Phys* 72 (1980) 2384-2393.

[5] Hansmann U.H.E., Okamoto Y. "The generalized-ensemble approach for protein folding simulations". In: Stauffer D., editor. Ann Rev Comput Phys VI. Singapore: World Scientific; 1999. p 129-158.

[6] Mitsutake A., Sugita Y., Okamoto Y. "Generalized-ensemble algorithms for molecular simulations of biopolymers". *Biopolymers (Pept Sci)* 60 (2001) 96-123.

[7] Berg B.A. "Biased metropolis sampling for rugged free energy landscapes". *Comp Phys Commun* 104 (2002) 52-57.

[8] Okumura H. "Partial multicanonical algorithm for molecular dynamics and Monte Carlo simulations". *J Chem Phys* 129 (2008) 124116.

[9] Okumura H. "Optimization of partial multicanonical molecular dynamics simulations applied to an alanine dipeptide in explicit water solvent". *Phys Chem Chem Phys* 13 (2011) 114-126.

[10] Berg B.A, Neuhaus T. "Multicanonical algorithms for first order phase transitions". *Phys Lett B* 267 (1991) 249-253.

[11] Berg B.A, Neuhaus T. "Multicanonical ensemble: a new approach to simulate first-order phase transitions". *Phys Rev Lett* 68 (1992) 9-12.

[12] Hansmann U.H.E., Okamoto Y., Eisenmenger F. "Molecular dynamics, Langevin and hybrid Monte Carlo simulations in a multicanonical ensemble". *Chem Phys Lett* 259 (1996) 321-330.

[13] Nakajima N., Nakamura H., Kidera A. "Multicanonical ensemble generated by molecular dynamics simulation for enhanced conformational sampling of peptides". *J Phys Chem B* 101 (1997) 817-824.

[14] Okumura H., Okamoto Y. "Monte Carlo simulations in multibaric-multithermal ensemble". *Chem Phys Lett* 383 (2004) 391-396.

[15] Okumura H., Okamoto Y. "Monte Carlo simulations in generalized isobaric-isothermal ensembles". *Phys Rev E* 70 (2004) 026702.

[16] Okumura H., Okamoto Y. "Liquid-gas phase transitions studied by multibaric-multithermal Monte Carlo simulations". *J Phys Soc Jpn* 73 (2004) 3304-3311.

[17] Okumura H., Okamoto Y. "Molecular dynamics simulations in the multibaric-multithermal ensemble". *Chem Phys Lett* 391 (2004) 248-253.

[18] Okumura H., Okamoto Y. "Multibaric-multithermal ensemble molecular dynamics simulations". *J Comput Chem* 27 (2006) 379-395.

[19] Ferrenberg A.M., Swendsen R.H. "Optimized Monte Carlo data analysis". *Phys Rev Lett* 61 (1988) 2635-2638. ibid. 63 (1989) 1658(E).

[20] Berg B.A., Celik T. "New approach to spin-glass simulations". *Phys Rev Lett* 69 (1992) 2292-2295.

[21] Lee J. "New Monte Carlo algorithm: Entropic sampling". *Phys Rev Lett* 71 (1993) 211-214. ibid. 71 (1993) 2353(E).

[22] Okamoto Y., Hansmann U.H.E. "Thermodynamics of helix-coil transitions studied by multicanonical algorithms". *J Phys Chem* 99 (1995) 11276-11287.

[23] Wang F., Landau D.P. "Efficient, multiple-range random walk algorithm to calculate the density of states". *Phys Rev Lett* 86 (2001) 2050-2053.

[24] Okumura H., Okamoto Y. "Multibaric-multithermal molecular dynamics simulation of alanine dipeptide in explicit water". *Bull Chem Soc Jpn* 80 (2007) 1114-1123.

[25] Okumura H., Okamoto Y. "Temperature and pressure dependence of alanine dipeptide studied by multibaric-multithermal molecular dynamics simulations". *J Phys Chem B* 112 (2008) 12038-12049.

[26] Sayle R.A., Milner-White E.J. "RASMOL: biomolecular graphics for all". *Trends Biochem Sci* 20 (1995) 374-376.

[27] Kollman P.A., Dixon R., Cornell W., Fox T., Chipot C., and Pohorille A. in *Computer Simulation of Biomolecular Systems* 3 A. Wilkinson, P. Weiner, and W. F. van Gunsteren (Eds.) (Elsevier, Dordrecht, 1997) p. 83.

[28] Jorgensen W.L., Chandrasekhar J., Madura J.D., Impey R.W., Klein M.L. "Comparison of simple potential functions for simulating liquid water". *J Chem Phys* 79 (1983) 926-925.

[29] Yoshida H. "Construction of higher order symplectic integrators". *Phys Lett A* 150 (1990) 262-268.

[30] Miller T.F., Eleftheriou M., Pattnaik P., Ndirango A., Newns D., Martyna G.J. "Symplectic quaternion scheme for biophysical molecular dynamics". *J Chem Phys* 116 (2002) 8649-8659.

[31] Tuckerman M., Berne B.J., Martyna G.J. "Reversible multiple time scale molecular dynamics". *J Chem Phys* 97 (1992) 1990-2001.

[32] Okumura H, Itoh S.G., Okamoto Y. "Explicit symplectic integrators of molecular dynamics algorithms for rigid-body molecules in the canonical, isobaric-isothermal, and related ensembles". *J Chem Phys* 126 (2007) 084103.

[33] Takekiyo T., Imai T., Kato M., Taniguchi Y. "Temperature and Pressure Effects on Cnformatonal Equilibria of Alanine Dipeptide in Aqueous Solution". *Biopolymers* 73 (2004) 283-290.

[34] Okumura H. "Temperature and pressure denaturation of chignolin: Folding and unfolding simulation by multibaric-multithermal molecular dynamics method". *Proteins* 80 (2012) 2397-2416.

[35] Honda S., Yamasaki K., Sawada Y., Morii H. "10 residue folded peptide designed by segment statistics". *Structure* 12 (2004) 1507-1518.

[36] Imamura H., Kato M. "Unfolding of β-hairpin peptides by pressure : FT-IR and FRET studies". *The 48th Annual Meeting of the Biophysical Society of Japan (2010)* 3P066.

[37] Hornak V., Abel R., Okur A., Strockbine B., Roitberg A., Simmerling C. "Comparison of multiple amber force fields and development of improved protein backbone parameters". *Proteins* 65 (2006) 712-725.

In: Models in Bioscience and Materials Research ISBN: 978-1-62808-052-0
Editor: Kh. T. Kholmurodov

Chapter 7

PROBING FREE ENERGY OF TRANSMEMBRANE HELIX-HELIX ASSOCIATION VIA A NOVEL SINGLE-RESIDUE BASED APPROACH

***Andrey S. Kuznetsov*[1,2,*], *Pavel E. Volynsky*[1] and *Roman G. Efremov*[1,2]**
[1] M.M. Shemyakin and Yu.A. Ovchinnikov Institute of Bioorganic Chemistry , Russian Academy of Sciences, Moscow, Russia
[2] Moscow Institute of Physics and Technology (State University), Moscow Region, Russia

Abstract

Free energy profiles for each type of interaction can be obtained based on a given spatial structure of the transmembrane (TM) dimer using molecular dynamics (MD) with umbrella sampling and potential of mean force (PMF) calculation. Glycophorin A (GpA) is one of the well-studied membrane proteins. There is a lot of experimental and theoretical data obtained, but little is known about the role of environment in GpA TM dimer formation. So, the main goal of the present work is to estimate contributions of all components (protein, lipids, water) in association of TM helices for the wild-type GpA TM dimer and its mutant form. Also, a new algorithm of residue-based energy decomposition was proposed and tested.

[*]E-mail address: andrej.kuznecov@phystech.edu

Keywords: Free Energy Calculations, Transmembrane Helix-Helix Association, Single-Residue Based Approach, Residue-Based Energy Decomposition

1. Introduction

Transmembrane (TM) alpha-helical segments are the basic structural units for a variety of membrane proteins. In case of multiple-span membrane proteins their association results in three-dimensional (3D) structure formation, while for single-span membrane proteins it plays an important functional role. Oligomerization of membrane receptors is thought to be very important for signal transduction, especially for single-span proteins like receptor tyrosine kinases (RTKs). In this case, conformational rearrangement in TM region is the only way to transmit signal from extracellular domain to cytoplasmic tyrosine kinase domain. It is well known that the process of TM helix-helix association is governed by their amino acid sequences [1], lipid composition of membrane [2] and other external factors. Mutations in TM domains can lead to receptor malfunction and corresponding diseases, but it is not clear yet how small single-residue changes can result in great changes in receptor activity [3, 4]. There are two properties that can be affected by the single point mutations: 3D structure of TM domain and free energy of association that determines the ability of sequences to associate [3]. These intuitive terms conceal a large number of complex physical processes that take place during association. One can say that changes in sequence can directly alter protein-protein interactions, but the role of the environment remains underestimated. It is convenient to study the behavior of protein in membrane-mimetic systems such as micelles or bicelles [5]. Unfortunately, they consist of a few types of lipid molecules, mostly only one specific lipid is used, but cell membranes have many components and exhibit lateral heterogeneity that results in formation of lipid microdomains with various physical properties [6, 7]. These factors can also modulate TM helices structure and oligomerization mechanism and should be taken into account in studies of protein-protein interactions in membranes.

Spatial structure of TM helical dimers can be predicted by computational methods [2, 8, 9] or can be determined using complex experimental procedures like NMR spectroscopy in membrane mimics (detergent micelles or bicelles) [5]. In a given environment one can obtain a single 3D structure by NMR or a number of structures predicted by molecular simulation techniques [2, 10]. But the information about the role of local membrane properties and tendency of a given helix to associate in vivo can be omitted in this approach. On the other hand, it can be crucial for the development of potent selective molecules targeting TM regions and modulating association of RTKs or other membrane receptors. There are several techniques for experimental estimation of association free energy: ToxR bacterial reporter systems (TOXCAT) [4, 11], Förster resonance energy transfer (FRET) [12] and the steric trap [13] – but the resulting values can vary in a wide range depending on the technique. Another problem is that all these methods say nothing about potential role of lipids in the dimerization process. They give only one total estimation value that cannot be translated into component contributions. In contrast, computer simulations can succeed in solving such a problem [3, 14]. Free energy profiles for each type of interaction can be obtained based on a given spatial structure of the dimer using molecular dynamics (MD) with umbrella sampling and potential of mean force (PMF) calculation [15]. Glycophorin A (GpA) is one of the well-studied membrane proteins. There is a lot of experimental and theoretical data obtained, but little is known about the role of environment in GpA TM dimer formation. So, the main goal of the present work is to estimate contributions of all components (protein, lipids, water) in association of TM helices for the wild-type GpA TM dimer and its mutant form. Also, a new algorithm of residue-based energy decomposition was proposed and tested.

2. Materials and Methods

2.1. MD Calculations

All MD simulations were performed using the GROMACS 4.5.2 package [16] and the GROMOS96 force field (43a2x parameter set). This was done with an integration time step of 2 fs, using 3D periodic boundary conditions, in the isothermalisobaric (NPT) ensemble with a semiisotropic (for relaxation) or isotropic (for production runs) pressure of 1 bar and a constant temperature of 315 K (to ensure the liquid crystalline phase of the lipid bilayers). The pressure and the temperature were controlled using the V-rescale thermostat and Berendsen barostat with 0.1 and 10.0 ps relaxation parameters, respectively, and a compressibility of $4.5 \cdot 10^{-5}$ bar^{-1} for the barostat. A twin-range (10/12 Å) spherical cutoff function was used to truncate van-der-Waals interactions. Electrostatic interactions were treated using the particle-mesh Ewald summation (real space cutoff 10 Å and 1.2 Å grid with fourth-order spline interpolation). Initial dimer model was based on the structure from the Protein Data Bank for the wild-type GpA (2KPF [17]). Mutant model was generated by homology (replacing corresponding residues) followed by MD-relaxation. The structures were truncated at the N- and C-termini preserving two charged residues on both ends to keep their correct positioning in a membrane

(sequences under study: SEPEITLIIFGVMAGVIG[T,V]ILLISYGIRR)

and then were inserted in pre-equilibrated palmitoyl-oleoyl phosphatidylcholine (POPC) bilayer of 128 lipids solvated by SPC water. Water molecules located in the bilayer interior were removed. In the beginning of the simulations, the systems were equilibrated by energy relaxation ($5 \cdot 10^4$ steps of steepest descent minimization) followed by heating from 5 K to the temperature of simulations (315 K) during 50-ps MD run and 10-ns MD run at 315 K with fixed positions of the peptides' atoms. Such equilibration is necessary for minimization of bilayer distortion induced by insertion of the peptides and removal of several lipid molecules. After that long-term (50 ns) collection MD runs were carried out.

2.2. Energy Profiles Calculations

PMFs were calculated as a function of helix-helix distance. In total, 66 windows were taken for umbrella sampling from $r = 7$ Å to $r = 20$ Å. The distance between centers of mass was considered as a reaction coordinate r. A set of reference peptides' structures were obtained by translating monomers in the membrane plane along the line connected their centers of mass. Initial structure for each window was generated using 10 ns MD equilibration (parameters similar to those mentioned before) when the monomers' positions were fitted to the target reference structure with a given distance between them. The 16 ns production MD runs were carried out in each window with a harmonic force constant of 1000 kJ mol^{-1} nm^{-2}. First one nanosecond trajectory of these runs was treated as final equilibration and was not included in subsequent analysis. Total PMF of the dimer formation was calculated by integrating the mean force $\langle Fr(r) \rangle$, where Fr is the projection of total force on the distance between monomers, r [15]. Error estimation was done by dividing MD trace into 3 parts, which were used for independent PMF calculations.

2.3. Energy Decomposition

Energy decomposition over the interaction types was done using the following scheme:

1. reduction of the initial trajectory to one containing only parts of the system (protein; protein+lipids; protein+solvent);
2. recalculation of forces from coordinates;
3. averaging forces acting on monomers, projection on r;
4. integration;
5. subtraction of protein-protein PMF to obtain the contribution of other components (if needed).

Unfortunately, this procedure induces an increase of errors in decomposition profiles, but their shapes and minimum position remain the same. Energy decomposition by atomic groups was performed according to another scheme:

1. getting all forces acting on protein atoms from analysis of MD trajectory;
2. summation of forces over a group of interest;
3. averaging forces and calculation of projection on r;
4. subtraction of intermonomer forces;
5. integration.

3. Results and Discussion

In the present work we selected the following systems to study: two TM alpha-helices (dimer) of the wild-type (WT) glycophorin A (GpA) and its weak mutant form Thr87Val in POPC bilayer. This system is widely studied in experimental and computational works [3, 4]. There are several points of view on the role of Thr87 in GpA dimerization. It can be involved in hydrogen bond formation between monomers [13], but that is not always the case because there is no evidence for hydrogen bonding for GpA in detergent micelles [11]. Thr87 to Val mutation is disruptive [4, 11] because valine residue cannot form such a hydrogen bond, but it does not prevent oligomerization at all. So, we suggest that such a mutant is well-suited for development and testing of the new decomposition technique. The algorithms used to calculate free energy profiles for TM helices have been described in a number of studies [3, 4], but decomposition of the total free energy into contributions of individual residues has been never

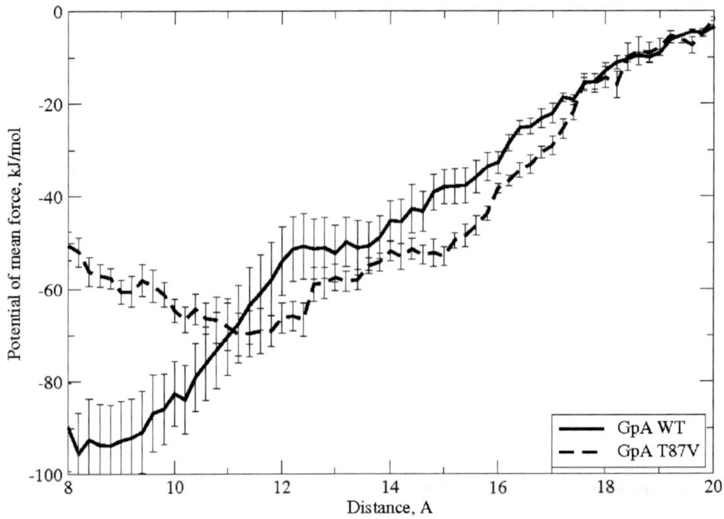

Figure 1. Total free energy profiles for wild-type glycophorin A (solid line) and T87V mutant (dashed line).

done before. We propose that Thr87Val mutation induces a slight unfavorable shift of the calculated PMFs. Moreover, this mutation shouldn't dramatically change 3D packing of the dimer because of approximately equal volume of the corresponding amino acids. So, we decided to skip the complicated structure prediction stage [2] by simple replacement of amino acid in the WT model followed by energy minimization.

As expected, there was no significant alternations in mutant dimer structure during MD run when compared to WT GpA. It was shown that the root-mean-square deviation (RMSD) between backbones of these models does not exceed 2.5 Å whereas dynamic changes in the course of one MD simulation of the dimer result in the structure variation within RMSD 2.2 Å from the starting state. The calculated total PMFs are shown in Figure 1. The first results obtained using the method reveal that the Thr87Val mutant has a less prominent minimum in the PMF profile than the WT dimer and it is located near the non-bonded state ($r_{min} = 11,5$ Å). To get additional insight into the dimerization mechanism, the total free energy of association was then decomposed into protein-protein and protein-environment contributions (Figure 2) that show how α-helices interact

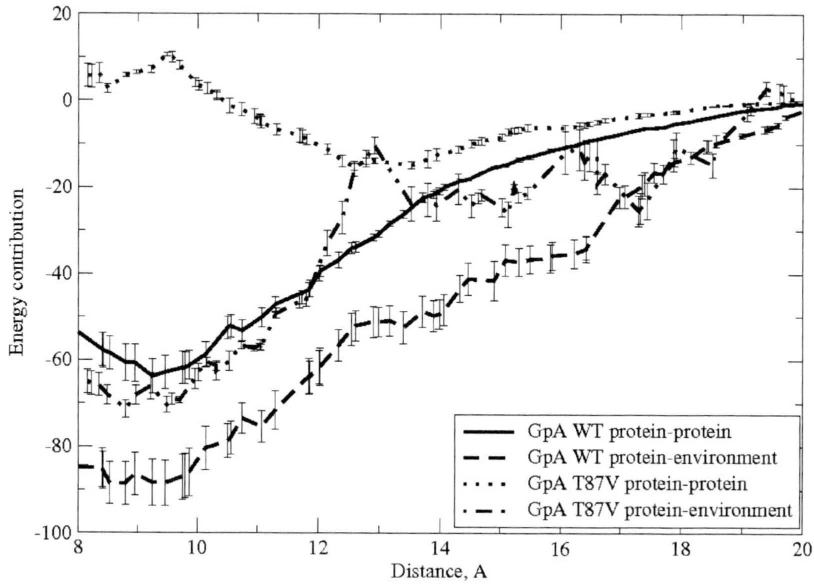

Figure 2. Free energy contributions from protein-protein and protein-environment interactions. Charged flanking residues were not included in calculation of protein-protein energy terms, so the corresponding pair sums may differ from the total graphs shown in Figure 1.

with each other and with lipids and water. As it was reasonable to expect, water molecules interact mostly with the flanking charged residues and there is no water contribution for residues in the middle part of the bilayer. The sums of lipids and water contributions are shown in Figure 2.

Finally, the contributions of individual residues into the binding energy were delineated (Figure 3). Surprisingly, the corresponding term for Val87 residue of the mutant was almost similar to that for Thr87 in the WT dimer (Figure 3). At the same time, this single mutation caused positive (unfavorable) shifts in energy contributions of the nearest interfacial residues (Gly83, Gly79). Also, we should mention that the non-interfacial residues were also affected by this mutation, but the general pattern of protein-environment interactions was more conservative. Noteworthy, the largest contribution into the free energy profiles was given by lipid-protein interactions that is in agreement with the results published by other groups [18]. In both cases protein-protein interactions seem to

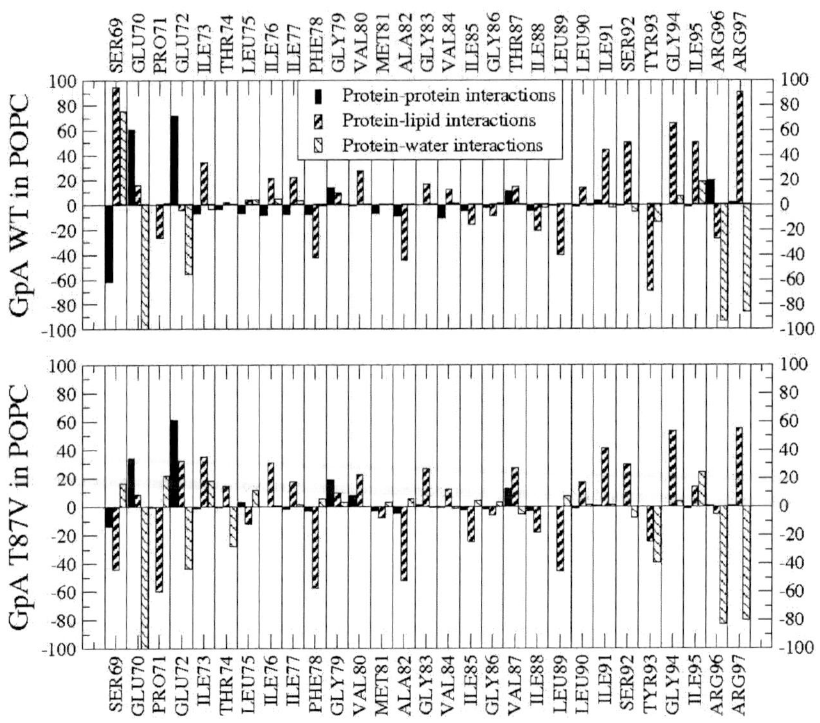

Figure 3. Free energy decomposition into residual contributions. All components of the energy are shown for GpA WT (top) and T87V (bottom).

play a minor role. As mentioned above, mutation Thr87Val strongly affects specific interactions between helices, but only slightly changes environmental contributions, so we can conclude that it will be weakly disruptive for the GpA association in lipid bilayers. Based on these observations, we suggest that the interaction with lipid environment can play a crucial role in association of TM helices. Further work is required to establish a proper balance of various factors governing helix-helix interactions in lipid bilayers.

Conclusion

In the present work, we introduced a novel computational approach to estimate how amino acid residues are involved in the process of helix-helix association in lipid bilayers. We have shown that in case of glycophorin A the major driving force of this process is lipid-protein interaction. Also, for a single mutation of interfacial residue we observed changes in energy contributions not only for the nearest residues, but also for distantly remoted non-interfacial ones. Therefore, we assume that the formation of TM dimers is a complex process, where cell membrane plays the main role at the first stage, thus bringing the helices together. By contrast, specific protein-protein interactions are important at the final stage of the dimer tuning. According to the hypothesis of multistate operation for membrane bitopic proteins [2], such a change can shift the equilibrium between dimer's states and finally results in receptor malfunction.

Acknowledgments

This work was supported by the Ministry of Education and Science of the Russian Federation (State contract 07.514.11.4127), by the Russian Foundation for Basic Research and by the RAS Programmes Molecular and cellular biology and Basic fundamental research for nanotechnologies and nanomaterials. Access to computational facilities of the Joint Supercomputer Center RAS (Moscow) and Moscow Institute of Physics and Technology is gratefully acknowledged.

References

[1] Lemmon, M.A., Flanagan, J.M., Treutlein, H.R., Zhang, J., Engelman, D.M. Sequence specificity in the dimerization of transmembrane -helices. *Biochemistry*, 31, (1992), 12719-12725.

[2] Polyansky A.A., Volynsky P.E., Efremov R.G. Multistate Organization of Transmembrane Helical Protein Dimers Governed by the Host Membrane. *J. Am. Chem. Soc.*, 134, (2012), 1439014400.

[3] Taylor M.S., Fung H.K., Rajgaria R., Filizola M., Weinstein H., Floudas C.A. Mutations affecting the oligomerization interface of G-protein-coupled receptors revealed by a novel de novo protein design framework. *Biophys. J.*, 94, (2008), 24702481.

[4] Duong M.T., Jaszewski T.M., Fleming K.J., MacKenzie K.R. Changes in apparent free energy of helix-helix dimerization in a biological membrane due to point mutations. *J. Mol. Biol.* 371, (2007), 422434.

[5] Bocharov E.V., Volynsky P.E., Pavlov K.V., Efremov R.G., Arseniev A.S. Structure elucidation of dimeric transmembrane domains of bitopic proteins. *Cell Adh. Migr.* 4, (2010), 284-298.

[6] Lingwood D., Simons K., Lipid rafts as a membrane-organizing principle. *Science* 327, (2010), 4650.

[7] Pyrkova D.V., Tarasova N.K., Krylov N.A., Nolde D.E., Pentkovsky V.M., Efremov R.G. Dynamic clustering of lipids in hydrated two-component membranes: results of computer modeling and putative biological impact. *J. Biomol. Struct. & Dyn.* 31, (2013), 87-95.

[8] Vereshaga Ya.A., Volynsky P.E., Nolde D.E., Arseniev A.S., Efremov R.G. Helix interactions in membranes: lessons from unrestrained Monte Carlo simulations. *J. Chem. Theory* & Comput. 1, (2005), 1252-1264.

[9] Vereshaga Ya.A., Volynsky P.E., Pustovalova Ju.E., Nolde D.E., Arseniev A.S., Efremov R.G. Specificity of helix packing in transmembrane dimer of the cell death factor BNIP3: a molecular modeling study. *Proteins: Structure, Function, and Bioinformatics.* 69, (2007), 309-325.

[10] Efremov R.G., Vereshaga Ya.A., Volynsky P.E., Nolde D.E., Arseniev A.S. Association of transmembrane helices: what determines assembling of a dimer? *J. Comput. Aid. Mol. Design* 20, (2006), 27-45.

[11] Doura A.K., Kobus F.J., Dubrovsky L., Hibbard E, Fleming K.G. Sequence context modulates the stability of a GxxxG-mediated transmembrane helixhelix dimer. *J. Mol. Biol.* 341, (2004), 991998.

[12] Duneau J.P., Vegh A.P., Sturgis J.N. A dimerization hierarchy in the transmembrane domains of the HER receptor family. *Biochemistry* 46, (2007) 2010-2019.

[13] Hong H, Blois T.M., Cao Z., James U. Bowie J.U. Method to measure strong proteinprotein interactions in lipid bilayers using a steric trap. *PNAS* 107, (2010), 1980219807.

[14] Polyansky A.A., Volynsky P.E., Efremov R.G. Structural, dynamic, and functional aspects of helix association in membranes: a computational view. Adv. *Protein Chem. Struct. Biol.* 8, (2011), 129-161.

[15] Henin J., Pohorille A., Chipot C. Insights into the recognition and association of transmembrane -helices. The free energy of -helix dimerization in glycophorin A. *J. Am. Chem. Soc.*, 127, (2005), 8478-8484.

[16] Van der Spoel D., Lindahl E., Hess B., Groenhof G., Mark A.E., Berendsen H.J. GROMACS: fast, flexible, and free. *J. Comput. Che*m. 26, (2005), 1701-1718.

[17] Mineev K.S., Bocharov E.V., Volynsky P.E., Goncharuk M.V., Tkach E.N., Ermolyuk Y.S., Schulga A.A., Chupin V.V., Maslennikov I.V., Efremov R.G., Arseniev A.S. Dimeric structure of the transmembrane domain of glycophorin A in lipidic and detergent environments. *Acta Naturae*, 3, (2011), 90-98.

[18] Zhang J., Lazaridis T. Transmembrane helix association affinity can be modulated by flanking and noninterfacial residues. *Biophys. J.* 96, (2009), 44184427.

In: Models in Bioscience and Materials Research ISBN: 978-1-62808-052-0
Editor: Kh. T. Kholmurodov

Chapter 8

HYDROGEN-BOND NETWORKS AS STRUCTURE-DETERMINING CONJUGATED SYSTEMS

Yulia V. Novakovskaya*
Chemistry Department, Moscow State University
Leninskie Gory, Moscow, Russia

Abstract

Analysis of the electron density distribution in clusters composed of hydrogen fluoride and water molecules reveals the existence of both σ- and π-binding between molecules. The σ-kind density distribution determines the mutual orientation of molecules. A π-system may be delocalized conjugated, which provides additional stabilization of molecular clusters. HF_2^- and $H_5O_2^+$ ions are characterized by quasi-triple bonds between the electronegative atoms. The most long-lived species stabilized by π-binding are structural rings and closed or open hoops and strips composed of fused rings. It is conjugated quasi-aromatic π systems that determine the cooperativity phenomenon. Spatial extension and structure of conjugated π systems determine the typical features of liquid hydrogen fluoride and water.

Keywords: Hydrogen Bonds, Clusters, Hydrogen Fluoride, Water, Electron Density, Cluster Orbitals, Intermolecular σ- and π-Bonds, Conjugation

*E-mail address: jvnovakovskaya@gmail.com

1. Introduction

The idea about the existence of some special kind of intermolecular bonds between water or ammonia molecules is already a century-old [1]. At first, T.S. Moor and T.F. Winmill spoke only about a weak union formed by the molecules [1]. In the next quarter of the century, due to the works of M.L. Huggins [2], W.M. Latimer and W.H. Rodebush [3], and especially L. Pauling [4], the concept of hydrogen bond was developed. The amount of papers published in all languages for the last three quarters of a century would be impossible to estimate. It might even exceed a quarter of a million.

Hydrogen-bonded ensembles are present nearly everywhere. All biological processes proceed in H-bonded environments and involve molecular structures with H-bonds. Plants and all living organisms are composed of water, up to 90–99%. The first principal discovery in the field was the tetrahedral arrangement of H-bonded water molecules determined by their peculiar spatial charge distribution [5]. Later on, α-helices and β-sheets were recognized as the main H-bond stabilized protein forms [6], and the double helical structure of DNA molecules was found to be determined by interhelical hydrogen bridges [7]. Insofar as the majority of technological processes proceed in aqueous solutions, and, which is more important, modern technology attempts to mimic the nature on the micro- and nano-scales, we can say that most of the technological schemes are also determined by peculiar properties of H-bonded substances. Thus, everything in and around us is related one way or another to the properties of H-bonded systems, but what is the main reason for the unique role of water as a protonic solvent? Let us clarify the point.

2. Hydrogen Bonds: Conventional View

According to the novel IUPAC definition, which inherited much from numerous previous definitions, "the hydrogen bond is an attractive interaction between a hydrogen atom from a molecule or a molecular fragment X–H in which X is more electronegative than H, and an atom or a group of atoms in the same or a different molecule, in which there is evidence of bond formation" [8]. Here, X–H is the bond donor, while Y or Y–Z is the acceptor with an electron rich region, such as a lone pair of Y or a π-bonded electron pair of Y–Z.

The definition is accompanied by six criteria and six characteristics of hydrogen bonds. It is pointed that the formation of a hydrogen bond is determined primarily by electrostatic forces and dispersion interaction, as well as partial charge transfer between the donor and acceptor. Distinguishing the forces implies the perturbative nature of the interaction, which is reasonable in the case of typical intermolecular bonds of van-der-Waals kind, when two approaching particles slightly affect the electron density distributions of each other, and the interaction can be estimated as a sum of the first- and second-order perturbation-theory corrections (electrostatic and polarization energy respectively) to the zero-order value that corresponds to non-interacting particles. However, being agglomerated in H-bonded clusters or ensembles, molecules no longer can be treated as independent individual particles, since they acquire novel properties; and hydrogen bonds formed (in most cases) have more in common with ordinary chemical bonds than with van-der-Waals bonds.

This similarity is actually reflected in a number of characteristics that accompany the definition of the bond [8], but is not explicitly formulated. For example, in terms of Bader's atoms-in-molecules analysis, electron density distribution in the X–H... Y–Z domain is usually characterized by a bond path between H and Y with a (3, –1) critical point typical of chemical bonds. Next, X–H... Y–Z hydrogen bond has characteristic NMR signatures including the proton deshielding, spin-spin coupling between X and Y, and nuclear Overhauser enhancements, which is usually observed in the situations when an X... Y bond exists. Furthermore, the phenomenon of cooperativity manifested in deviations of the properties from pair-wise additivity is typical of hydrogen-bond networks. Finally, there are geometrical features of hydrogen bonds (such as directionality of the bonds and the closeness of X–H... Y angles to 180°) and the energetic features (such as the increase in the bond strength with an increase in the electronegativity of X and an increase in the X–H distance accompanied by the decrease in the X–H stretching frequency), which can be determined only by certain changes in the electron density distribution in H... Y domains untypical of ordinary intermolecular bonds.

Then, the following question arises: what is the kind of electron binding that determines the formation of hydrogen bonds? Particularly, what is the reason for the extra strong bonds observed in $H_5O_2^+$ (36 kcal/mol) and HF_2^- (40 kcal/mol) systems, which have symmetric configurations where a proton is located accurately at the midpoint between two oxygen or fluorine nuclei so that the structures are stabilized by two equivalent H... X bonds (X = F or O) rather than one covalent and one hydrogen bonds (X–H... X)?

Unfortunately, the recent definition of hydrogen bond, in which all the characteristics of the bonds are embraced and systematized, did not clarify the nature of the bonds. Let us try to do that.

3. Separate Hydrogen Bonds Versus Hydrogen Bond Networks

To start with, let us distinguish two obvious kinds of hydrogen bonds. In the case of molecules, which involve FH, OH and NH groups (primarily, hydrogen fluoride, water, and ammonia), H-bonded networks can appear. At the same time, there are examples of H-bonded molecular pairs that cannot be joined with each other. Therefore, it is reasonable to speak about two kinds of hydrogen bonds: (i) separate bonds formed each by two particles and (ii) bond networks that unite a large number of molecules. The weakest bonds are typically of the former kind, while the strongest bonds are of the latter kind.

Bond networks appear when each involved particle can form no less than two H-bonds, one as a proton donor (*d*) and the other as an acceptor (*a*). This is possible when there are a proton, which is chemically bound to an electronegative atom, and a "free" valence electron density (uninvolved in interatomic bonds). When hydrogen bonds and conventional intramolecular bonds are alternating (X–H...X–H...X–H...), a system of interacting bonds rather than just a spatial bond network originates, and the molecules manifest collective properties. Below, we analyze electronic structures of the clusters composed of water and hydrogen fluoride molecules in order to clarify the character of the electron density redistribution that determines the properties of inherently consistent molecular ensembles.

4. Features of the Electron Density Distribution in Hydrogen-Bonded Clusters

Approximation of the electron density $\rho(\mathbf{r})$, which is most convenient for the analysis, is the expansion in the basis of natural orbitals ($\phi_i(\mathbf{r})$):

$$\rho(\mathbf{r}) = \sum_{i=1}^{P} n_i |\phi_i(\mathbf{r})|^2,$$

where n_i coefficients are effective occupancies of the orbitals. When the coefficients are close to one or two (i.e., negligibly differ from these integers), one can believe (just believe, since this cannot be proved) that $|\phi_i(\mathbf{r})|^2$ functions, which were found as a result of nonempirical calculations, represent reasonable approximations of the actual charge distributions of individual electrons.

In the case of water and hydrogen fluoride molecules (in the ground electronic states), as well as clusters composed of these molecules, ni numbers in the above equation meet this requirement; and natural orbitals constructed in the second-order of the Moeller-Plesset perturbation theory (MP2), as well as with the use of fully or partially variation schemes, such as single- or multi-reference configuration interaction method or coupled clusters approach, are very close to the solutions obtained at the single-determinant Hartree-Fock (HF) level.

Results considered below are obtained in quantum chemical calculations carried out at the MP2 level with a 6-31G basis set augmented with diffuse and polarization functions on all nuclei (6-31++G(d,p)). This level of theory was repeatedly proved to be adequate for describing the systems of interest. All the configurations correspond to the energy minima, which is confirmed by the normal-coordinate analysis. Dissociation energies are estimated taking into account counterpoise corrections for the basis set superposition error, and the mean H-bond energy is found by dividing the total dissociation energy of a cluster ($X_n \rightarrow nX$, where X = H_2O or HF) by the number of H-bonds. All calculations were carried out with the use of GAMESS [9] and Firefly [10] program packages.

When considering $|\phi_i(\mathbf{r})|^2$ contributions to the electron density distribution, let us use the following approach. The most straightforward physical interpretation of the quantum chemical description of the electronic subsystems of individual molecules is provided by the MO LCAO scheme (where molecular orbitals are approximated by linear combinations of atomic orbitals), which is founded when a single-determinant description of the electronic system provides reasonable results. Let us similarly consider natural orbitals of molecular clusters (cluster orbitals, CO) as linear combinations of molecular orbitals (LCMO). Such CO LCMO scheme is no less founded than conventional MO LCAO one, since relative changes in the electron density distribution of individual molecules that accompany the formation of intermolecular bonds are definitely smaller than those typical of intramolecular chemical bonds.

The sole substantial difference between the formation of intra- and intermolecular bonds is as follows. Intramolecular bonds form due to the overlap of the frontier orbitals of interacting particles (atoms or functional groups). These are both highest occupied and lowest unoccupied molecular orbitals or singly occupied molecular orbitals of both reagents; and in the resulting system (in the ground electronic state), bonding (and nonbonding) orbitals are usually occupied, while antibonding ones are unoccupied. When one considers intermolecular bonds within clusters composed of closed-shell molecules, the situation is different: the electron density redistribution is due to the overlap of the occupied orbitals of the particles, so that the resulting functions, which describe the electron density distribution in the cluster, are of both bonding and antibonding kinds. Hence, the total binding effect is much smaller compared to typical interatomic bonds and depends on the balance between bonding and antibonding contributions within each intermolecular region, particularly within H-bond domains.

Let us denote[1] orbitals of individual molecules (molecular orbitals) as ϕ_i^f and ϕ_i^w in the case of hydrogen fluoride and water respectively, while those of clusters (cluster orbitals) as φ_i^f and φ_i^w. Below, we analyze the LCMO expansions of the cluster orbitals of $(HF)_n$ and $(H_2O)_n$ systems, neglecting the strongly localized 1s core orbitals of fluorine and oxygen atoms (ϕ_1^f and ϕ_1^w molecular orbitals), which remain nearly unchanged upon the formation of clusters and, hence, do not contribute to intermolecular binding.

4.1. $(HF)_n$ Clusters

Let us briefly remind how the bonding orbitals of individual HF molecule (Fig. 1a)[2] look like omitting in the mathematical expressions those atomic orbitals, which contribute with negligible or much smaller weights, and dropping coefficients that depend on internuclear distances and particular functional form of the orbitals. The chemical bonding is determined by two orbitals, one of which (ϕ_2^f) is a linear combination of $2s$(F) and $1s$(H) atomic functions,[3] while the other (ϕ_3^f) is a combination of $2p_z$(F) and $1s$(H) functions (Oz axis is oriented along the F–H line). The residual four valence electrons are described by two degenerate orbitals, namely, $\phi_4^f = 2p_x(\mathrm{F})$ and $\phi_5^f = 2p_y(\mathrm{F})$, and represent the

[1] Sometimes, superscripts are omitted when no confusion may arise (e.g., in figures).

[2] Figures 1, 3, 4.1, and 4.2–4.3 show the surfaces of constant $\phi_i = 0.0015$ or $\varphi_i = 0.0015$ value, i.e., $\rho_i = 2.25 \cdot 10^{-6}$ au. The darker regions correspond to the positive values of ϕ_i or φ_i function, while the lighter regions, to the negative values, so that their boundaries are the zero electron density surfaces. Orbitals in Figs. 1 and 3 are shown as transparent so that one can clearly see positions of nuclei.

[3] We use conventional notations for the atomic orbitals that involve shell numbers.

electron density that can be referred to as "free", (uninvolved in interatomic binding) and supposedly determining intermolecular bonds.

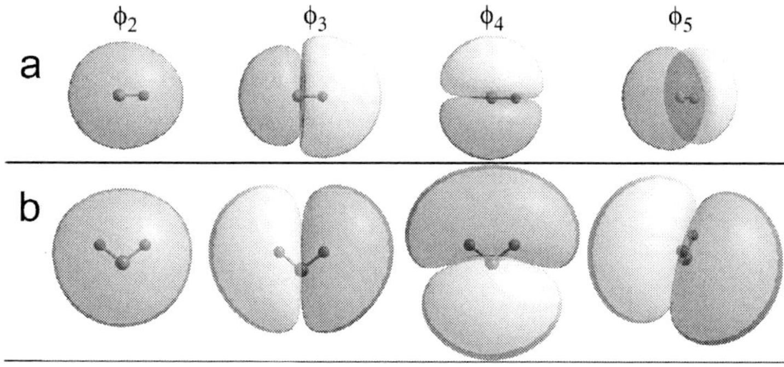

Figure 1. Valence molecular orbitals of individual (a) HF and (b) H_2O molecules.

HF molecule typically forms two bonds with the neighbors (*da* coordination), though there are situations when either hydrogen or fluorine atom or both of them are involved in two bonds with the neighboring molecules. Stable structural units are molecular rings (Fig. 2a) where molecules are arranged in such a way that H...F–H angles are close to 90°, while dihedral F–H...F–H angles are arbitrary. When no less than six molecules are joined in a ring, it looks like a folded sheet or strip composed of rectangular segments (Fig. 2b). Openwork species, in which three-membered molecular rings have common HF edges, are also possible (Fig. 2c). Larger rings that involve three-coordinated molecules can be joined in nonplanar openwork species. Such building blocks enable one to construct a two-dimensional network, but any attempt to build a continuous three-dimensional bonded system fails.

The mean H-bond energy of $(HF)_n$ rings (Fig. 2a) changes from 5.2 to 8.6 kcal/mol when the number of molecules increases from 3 to 5. In extended folded rings similar to that shown in Fig. 2b, the mean bond energy is about 8.4 kcal/mol. Two parallel molecular rings have no noticeable attraction to each other. Species based on trimolecular rings with common HF edges (Fig. 2c) are less stable than plain rings: a mean H-bond energy is about 4.2 kcal/mol. With an increase in the number of molecules, which form more than two hydrogen bonds, the mean bond energy decreases to about 4 kcal/mol. A mean bond energy per molecule is higher, about 5 kcal/mol, but still substantially lower than that in plain molecular rings. Bonds in chains are characterized by intermedi-

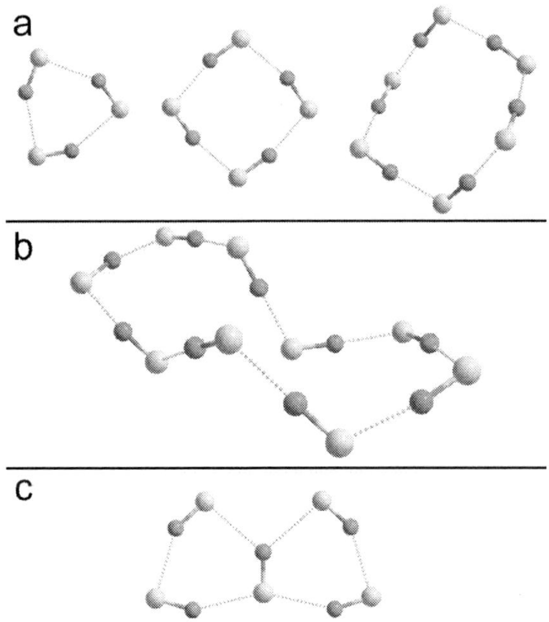

Figure 2. Typical structures of $(HF)_n$ clusters: (a) plain rings; (b) extended folded ring; and (c) openwork composed of trimolecular rings.

ate energies, e.g., about 5.9 and 6.4 kcal/mol in four- and five-membered chains. Note that the difference between the bond energies in molecular rings and chains (2–2.5 kcal/mol) seems too large for the systems composed of similarly coordinated molecules. Furthermore, when a zigzag tetra- or pentamolecular chain closes in a ring, the absolute energy decrease is 17–20 kcal/mol, which is nearly thrice as large as an individual H-bond energy in the ring. This means that such closing is accompanied by the additional stabilization of the already existing hydrogen bonds. Let us clarify the nature of the stabilization by analyzing the character of cluster orbitals.

Let us start with $(HF)_3$, which is the smallest ring-like structure. There are cluster orbitals of two kinds in the system (Fig. 3), namely of σ and π kinds. Let us denote the plane of fluorine nuclei as xOz; then, the former cluster orbitals (m = 4–9, 11, 14, 15) are linear combinations of ϕ_2^f, ϕ_3^f, and ϕ_4^f molecular orbitals respectively, while the latter (m = 10, 12, 13) are linear combinations of ϕ_5^f molecular orbitals and have much in common with typical π-system functions of conjugated hydrocarbons. In hydrocarbon ring skeletons, maxima of the π functions are at carbon nuclei, while minima or zeros are at the midpoints of C–C bonds. In $(HF)_3$ cluster, the large-value parts of orbitals correspond to H–F regions with the maxima at F nuclei and minima in H...F domains close to bridge proton positions. This means that π cluster orbitals describe a delocalized π-binding of fluorine atoms, and H atoms act here as auxiliary bridge sites. Note that in the regions of intramolecular H–F bonds, all σ and π cluster orbitals remain bonding but are either bonding or antibonding in the hydrogen-bond domains.

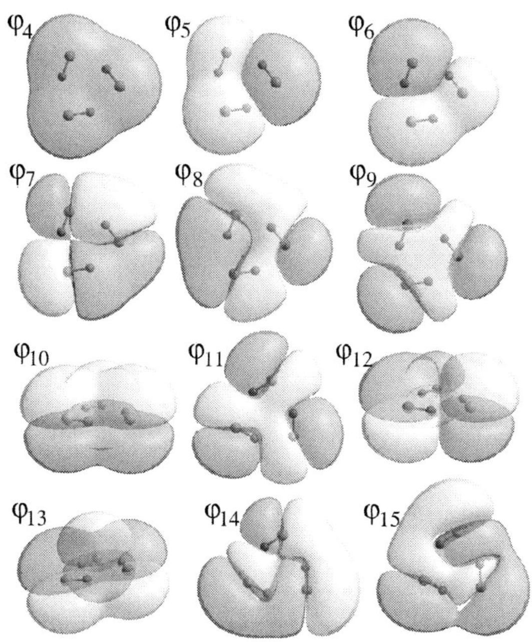

Figure 3. Valence orbitals of $(HF)_3$ cluster.

Insofar as σ cluster orbitals are linear combinations of either $\phi_2^f = 2s(\mathrm{F}) + 1s(\mathrm{H})$ or $\phi_3^f = 2p_z(\mathrm{F}) + 1s(\mathrm{H})$ bonding orbitals and $\phi_4^f = 2p_x(\mathrm{F})$ lone pair orbitals, whose axes are mutually orthogonal, HF molecules tend to be arranged in clusters in such a way that H...F–H angles are close to 90°. Planar configuration of $(HF)_3$ ring also seems natural, since the additional π binding is provided by the overlap of $\phi_5^f = 2p_y(\mathrm{F})$ molecular orbitals whose axes are normal to the plane of σ-binding.

The same character is typical of the four-membered $(HF)_4$ ring and openwork $(HF)_n$ clusters, in which trimolecular rings are joined via common HF edges (see Fig. 2). On the whole, changes in the orbital picture that take place with an increase in the ring size or upon fusion of the rings are similar to those observed in the case of hydrocarbon rings when the number of C nuclei in the ring skeleton increases or when rings of the same size are joined via edges. For example, five occupied π orbitals of $(HF)_5$ openwork structure have much in common with π orbitals of naphthalene.

The similarity of the orbital pictures of small $(HF)_n$ rings and aromatic hydrocarbons does not apply to chain-like structures. In planar molecular $(HF)_4$ or $(HF)_5$ chains, there are both σ and π cluster orbitals, but each π orbital describes chiefly the electron binding of some two neighboring molecules. By contrast to rings, π orbitals in the chains are noticeably localized. Then, the aforementioned substantial stabilization of the cluster upon the chain closure in a ring is due to the appearance of a delocalized conjugated quasi-aromatic π system.

To sum up, the combination of σ and π binding explains the relative stability of small planar $(HF)_n$ clusters. Planarity is the characteristic feature of conjugated hydrocarbons. However, as was mentioned above, when there is no less than six molecules in an $(HF)_n$ ring, it is no longer planar (even at n=5, there is already a small deviation from planarity). It is folded and consists of rectangular segments. This peculiarity agrees with the trend of HF molecules to form sequences of mutually orthogonal hydrogen bonds, which is impossible in the case of planar structures. The orthogonality is determined by σ binding. Hence, we can state that this kind of electronic binding is present in all the clusters and determines the spatial arrangement of molecules. Then, what is the role of π orbitals and do they exist in nonplanar systems? It turns out they do, but in some new variant.

As an example, let us consider π orbitals of a ring-like $(HF)_6$ cluster (Fig. 4.1a). The lowest bonding orbital of this kind (φ_a) has much in common with usual π orbitals of conjugated hydrocarbons. All the residual π cluster orbitals dramatically differ from those of conjugated hydrocarbons and planar $(HF)_n$ clusters. Due to the planar configuration of all the latter systems, there were definite charge-alternation surfaces in all of them located in the H-bond domains. In $(HF)_6$ cluster, constant-sign parts of the orbitals look as closed rings, and the rings are connected as links of a chain (φ_b in Fig. 4.1a). The antibonding regions are located either in the vicinity of bridge protons or in the central part of the cluster; and electrons of all six π orbitals, each of which is delocalized over the whole hexamolecular ring, contribute to the intermolecular binding.

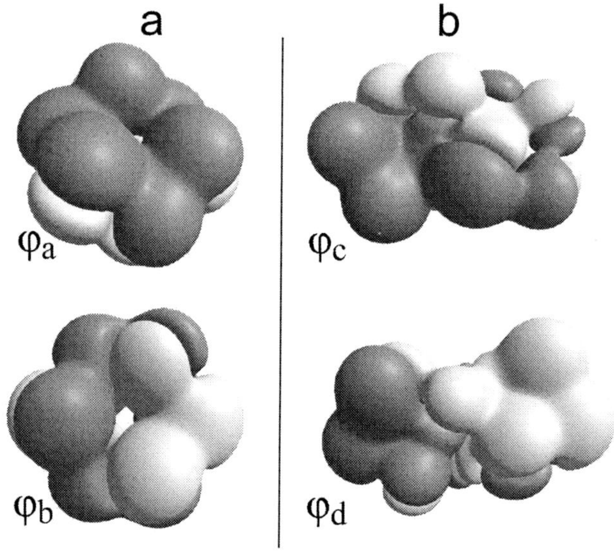

Figure 4. Typical valence π-orbitals of nonplanar $(HF)_n$ clusters: (a) $n = 6$ and (b) $n = 9$.

When the number of molecules is increased above eight, and pronounced folded structures appear, the orbital picture becomes more complex (Fig. 4.1b). Even the lowest π orbital looks already as two interlaced rings (of different signs) and resembles a typical π bonding orbital only in two thirds of the ring (φ_c). In the next orbitals, which are more and more complex interlacements of multifold bent rings, the sign-alternation regions already coincide with the H-bond domains, so that the electronic binding in certain H...F regions is weaker and, at the same time, there is no general pronounced binding of all the molecules.

In a general case, if we look at a folded structure from the side, we can see a zigzag; and the more strongly internally bound subsystems correspond to separate periods of the zigzag. For example, in $(HF)_9$, tetra- and pentamolecular subsystems (φ_d in Fig. 4.1b) can be distinguished. Quantitatively, this separation of quasi-aromatic segments in extended folded structures is reflected by the mean H-bond energy of the cluster, which decreases from 9.2 kcal/mol in $(HF)_6$ (a single quasi-aromatic system) to 8.3 kcal/mol in $(HF)_9$ (a combination of two quasi-aromatic subsystems).

The discovered peculiarities of electronic binding between HF molecules show that hydrogen fluoride should form structures composed of lamellar domains, the most stable building units of which should be planar or folded (depending on the molecular size) rings. In the rings, molecules are joined via electronic bonds, in which σ and π contributions are clearly distinguishable, and π binding system is delocalized conjugated. With an increase in the molecular size of the ring, which is accompanied by the formation of nonplanar structures, the H-bond energies increase (to a certain limit) due to the stronger π binding. When a ring becomes too large, and no effective conjugated bond system can be formed, strongly bound subsystems tend to separate, which should actually take place with an increase in temperature. These peculiarities explain why hydrogen fluoride is mobile and volatile liquid.

4.2. $(H_2O)_n$ Clusters

In an H_2O molecule (Fig. 1b), the lowest valence orbital (ϕ_2^w) is a linear combination of $2s(O)$, $1s(H^1)$, and $1s(H^2)$ atomic functions (where superscript enumerates protons). Two next orbitals are bonding in O–H^1, O–H^2 and $H^1 \ldots H^2$ domains respectively: $\phi_3^w = 2p_x(O) + (1s(H^1) - 1s(H^2))$ (xOz is the nuclear plane of the molecule, and Oz axis is the symmetry axis of the molecule) and $\phi_4^w = 2p_z(O) + (1s(H^1) + 1s(H^2))$. Finally, a lone pair, which represents the free density of the molecule, is described by $\phi_5^w = 2p_y(O)$ orbital.

Let us distinguish the following coordination kinds of water molecules in clusters: *da*, *dda*, *daa*, and *ddaa*. If all molecules are involved in trimolecular rings, then, practically irrespectively of their particular local neighborhoods (*da* in a ring-like trimer or *da* and *ddaa* in a pentamer where two trimolecular rings are joined via a common molecule, Fig. 4.2a), the mean H-bond energy is 5.0 kcal/mol. In $(H_2O)_n$ rings of the larger size (Fig. 4.2b) where H-bond angles are closer to 180° and all molecules have *da* coordination, the interaction energy per bond is 6.5, 6.8, and 6.9 kcal/mol at $n = 4$, 5, and 6 respectively. In $(H_2O)_6$ prism, $(H_2O)_8$ cube, and $(H_2O)_{12}$ cage, which are composed of fused three-, four-, five-, and six-membered molecular rings where tricoordinated molecules prevail (Fig. 4.2a, c), the mean H-bond energy depends on the steric factor, being 4.7, 5.7, and 6.0 kcal/mol, respectively. Thus, with an increase in the number of hydrogen bonds formed by the same molecule, the mean bond energy slightly decreases, while with an increase in the size of the ring, it increases, so that the

H-bond energy per molecule increases with an increase in the molecular size of the cluster, being 6.9, 8.5, and 8.6 kcal/mol in the aforementioned $(H_2O)_n$ clusters with $n = 6$, 8, and 12. Thus, by contrast to hydrogen fluoride, the mean bond energy remains sufficiently high when the coordination number of the molecule increases from two to four; and the H-bond energy per molecule even increases (up to a certain limit). Let us clarify the reason for that by analyzing the cluster orbitals.

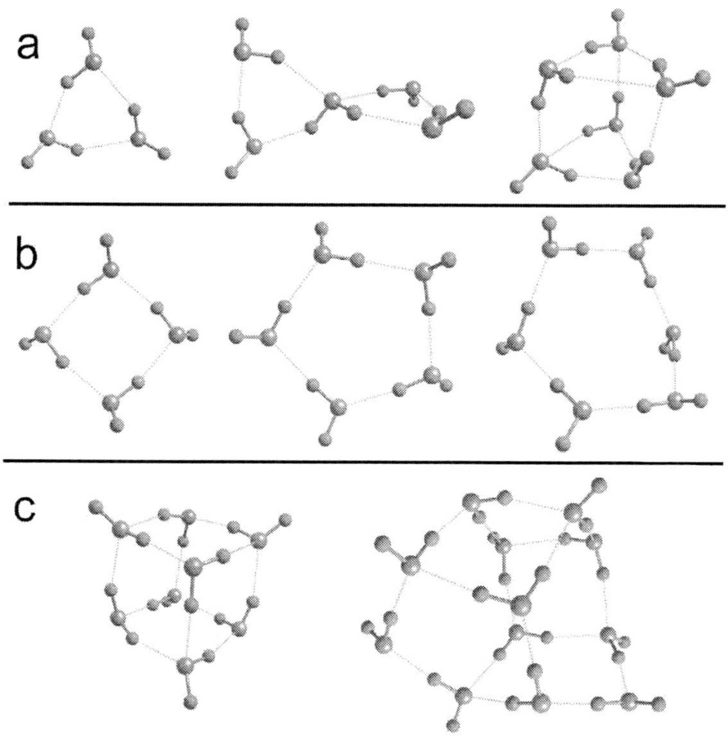

Figure 5. Typical structures of $(H_2O)_n$ clusters: (a) composed of or involving trimolecular rings; (b) plain rings with nearly straight hydrogen bonds; and (c) cages composed of fused tetra-, penta-, and hexamolecular rings.

Let us start (as before) with the smallest ring-like cluster, namely trimer. Recall that in $(HF)_3$, we could definitely distinguish σ and π cluster orbitals, and the latter orbitals were very similar to bonding and antibonding orbitals of a conjugated hydrocarbon. What do we see in the case of water trimer? Figure 4.2a shows three most interesting valence orbitals of the cluster. Their approximations involve ϕ_4^w and ϕ_5^w molecular orbitals and they look like slightly twisted π orbitals. They are twisted in such a way that within each O. . . H domain they have definitely π character. Alternation-sign surfaces of the orbitals always pass in close vicinity to bridge protons.

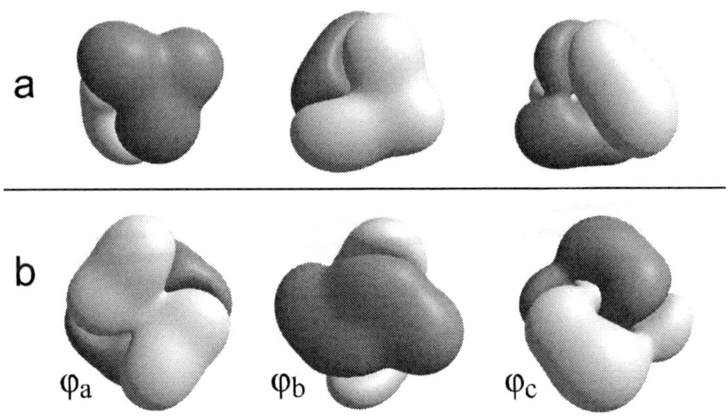

Figure 6. Some valence π orbitals of ring-like $(H_2O)_n$ clusters: (a) $n = 3$ and (b) $n = 4$.

The orbital picture of the ring-like water tetramer also represents an intermediate situation between those typical of planar and non-planar systems (Fig. 4.2b). There is an orbital, which resembles a nonbonding π orbital of a planar four-membered ring (φ_a). Another one (φ_b) is a π bonding orbital. Two residual orbitals are similar to those observed in non-planar clusters of hydrogen fluoride, namely, look like interlaced constant-sign rings (Fig. 4.2b shows one of them, φ_c), where the sign-alternation surfaces pass in close vicinity to bridge-proton positions. It is worth noting that, despite the visible resemblance, the orbitals have a definite peculiarity. Zero-density surfaces are very complex, and in many cases, they divide two bonding domains, namely, intramolecular O–H and the neighboring intermolecular H. . . O domains, so that the generally π-kind orbital that spans the whole cluster has a mixed π/σ character in the H-bond domains.

In ring-like pentamer and hexamer, the situation is principally similar. There are five and six cluster orbitals of generally π kind. Character of the electron density distribution described by any of the orbitals can be named intermediate between those typical of hydrogen fluoride clusters. This means that there are bonding and antibonding domains, but the constant-sign rings always look like bended or twisted doughnuts with tight constrictions. In fact, this observation strongly correlates with the "amount" of free electron density that can be localized in the H-bond domains. Accordingly, the difference in bond energies between hydrogen fluoride and water clusters increases with an increase in the number of molecules involved in the ring. It is about 0.1, 1.0, 1.6, and 2.4 kcal/mol in the case of trimers, tetramers, pentamers, and hexamers, respectively. Thus, rings composed of HF molecules are definitely more stable.

However, the situation dramatically changes when one goes to cage-like clusters. HF molecules do not agglomerate to form stable three-dimensional cages. By contrast, water molecules form three-dimensional networks. Mean H-bond energies in such systems are close to or even higher than 6 kcal/mol, and the bond energies per molecule are close to 8.5 kcal/mol, i.e., of the same order as in the most stable folded $(HF)_n$ clusters. The reason becomes clear if we consider Fig. 4.2 where orbitals of $(H_2O)_5$ cluster, which involves a tetracoordinated molecule, and those of $(H_2O)_8$ and $(H_2O)_{12}$ cages composed of fused molecular rings are shown. One of the orbitals of $(H_2O)_5$ (φ_a in Fig. 4.2a) looks like two concatenated H letters, one leg of each being normal to the residual T-shaped segment. Such orbital provides π-electron binding in the domains of all six H-bonds. Two other orbitals (nearly degenerate) look like concatenated twisted Z-shaped patterns (Fig. 4.2a shows one of them, φ_b) and provide π binding within one structural triangle and between two residual molecules. Similarly to the above orbitals of ring-like clusters, these orbitals have generally π character and a mixed π/σ character in the H-bond domains. Such flexibility of orbitals is determined by the contributions of ϕ_3^w, ϕ_4^w, and ϕ_5^w functions with different weights. Thus, a relatively small cluster that involves a tetracoordinated joint molecule is stabilized by π binding between all molecules.

Orbitals that span the whole or large part of the structure can be found also in $(H_2O)_8$ and $(H_2O)_{12}$ cages. Figure 4.2 (b, c) shows four such orbitals. In octamer, there are orbitals of diverse shapes, and here, the principally new situation is observed. First of all, there is an orbital (φ_c), which looks like a sandwich and describes conjugated π-binding systems of two four-membered rings that form two opposite faces of the cube (let us conditionally refer to them as the bottom and top). Another orbital of the same kind (φ_d) describes binding within four tetramolecular rings that form side faces of the cube. There are also orbitals that

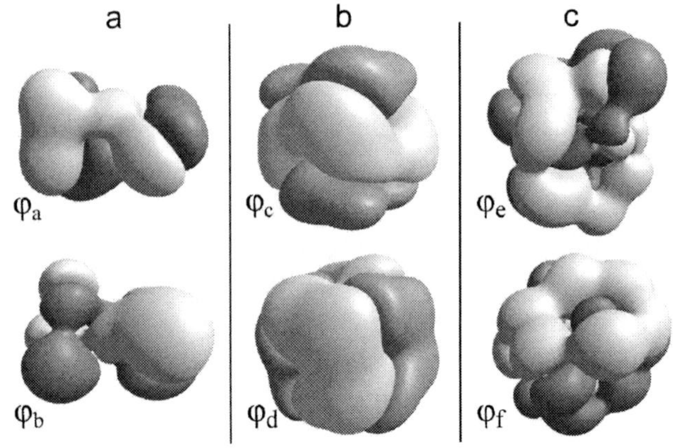

Figure 7. Some valence π orbitals of cage-like $(H_2O)_n$ clusters composed of fused or joined molecular rings and involving tri- and tetracoordinated water molecules: (a) $n = 5$, (b) $n = 8$, and (c) $n = 12$.

describe π binding in the domains of dimeric edges of the cube. Note that in all the cases, the orbitals have bonding or antibonding σ character in the directions normal to π bonds. This is the pronounced manifestation of the aforementioned flexibility of the electron density distribution in water clusters.

In the case of $(H_2O)_{12}$, let us consider also only two orbitals (φ_e and φ_f in Fig. 4.2c). At first glance, they are very similar to those of plain rings. In fact, they are. In one of them, interlaced constant-sign domains are composed of six- and five-membered rings connected via a typical π bonding segment of a tetramolecular ring. In the other orbital, two concatenated constant-sign patterns are based on a six-membered ring fused with five- and four-membered rings on opposite edges. As a result, there appears a conjugated π system that spans four structural rings, which in turn form a closed or open spatial hoop. Some other cluster orbitals look like interlaced or concatenated variously bent segments, and the corresponding electrons provide binding within four- to eight-molecular subsystems of the cluster.

These results explain why cage-like structures are more stable than rings. As soon as the number of molecules is sufficient for the formation of a cage composed of fused rings whose (mean) molecular planes are either nearly orthogonal to each other (as in small $(H_2O)_n$ clusters with $n = 6$ or 8) or form angles up to ca. 135° (as in $(H_2O)_n$ clusters with $n = 12$ or 20), such arrangement becomes preferable. A large number of cluster orbitals have mixed nature: they act as σ bonding (or antibonding) orbitals in one plane (ring) and as π bonding orbitals in the orthogonal subspace. It is worth noting that in this subspace, a continuous delocalized system, which spans several structural rings, is formed. The typical range of interplanar angles is determined by the mutual orientation of the initial molecular orbitals and their consistent contributions to σ and π binding. Moreover, even those orbitals, which look like antibonding in O–H... O region, have such spatial character that act as bonding in two subdomains, namely O–H and H... O.

Thus, even in cage-like clusters, there are delocalized conjugated π systems, which span either individual molecular rings or hoops and strips composed of fused rings. (Recall that in the case of hydrogen fluoride, a sequence of quasi-aromatic segments rather than a continuous system could only be formed.) Besides that, tetracoordinated water molecules can be effective joints between molecular rings in a three-dimensional network. Hence, diverse structural groups can be present in water ensembles as long-living structure units; and it is already clear that fused molecular rings involved in closed or open hoops should have the largest lifetimes among long-living species due to the general delocalized π-binding. As a result, water is a sole solvent that provides the formation of a continuous system of bonds, which is, on one hand, sufficiently flexible (due to the effective σ/π character of the electron density distribution within intermolecular domains) to form solvation shells of appropriate configurations around diverse particles and, on the other hand, quite strong (due to the conjugated π-electron system) to survive upon moderate external perturbations.

4.3. Ionic Clusters

Now, it is interesting to consider HF_2^- and $H_5O_2^+$ ionic clusters, which are known to be stabilized by the strongest hydrogen bonds.

In HF_2^-, the binding is provided by electrons described by a σ-orbital and two degenerate mutually orthogonal π orbitals (Fig. 4.3a shows one of them, φ_a). Moreover, zeros of two corresponding antibonding π orbitals accurately coincide with the proton position (Fig. 4.3a shows one of them, φ_b). As a result, judging from the shape of the orbitals, one can say that there is some kind of a triple bond between two fluorine atoms. Moreover, both bonding and antibonding orbitals describe electrons that provide binding in F... H domains.

The quasi-triple F... F bond and the nearly zero electron density of anti-

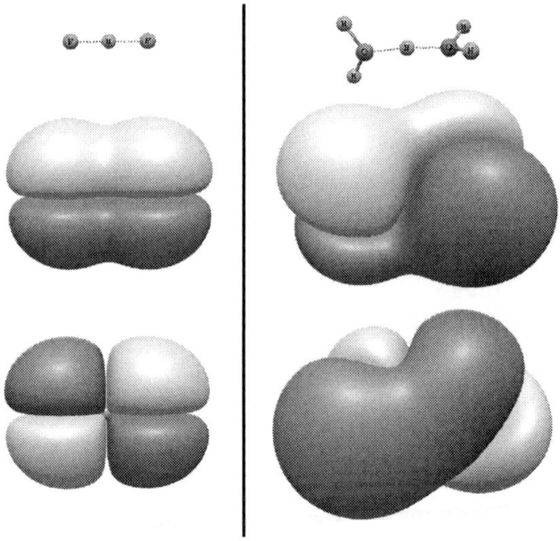

Figure 8. Typical π orbitals of (a) HF_2^- and (b) $H_5O_2^+$.

bonding orbitals in the vicinity of the bridge proton represent the limiting variant of the effects, which are less pronounced in neutral clusters. However, it is the effects that clearly explain the experimental NMR signatures of hydrogen bonds, namely, proton deshielding and spin-spin coupling of the nuclei of electronegative atoms.

A similar picture (though more complex) is observed in the case of $H_5O_2^+$. In this ion, there are five cluster orbitals, each of which looks like two embraced constant-sign segments (two of them are shown in Fig. 4.3b, φ_c and φ_d). Accordingly, each orbital describes electronic binding of two H_2O fragments (with the zeros close to the position of the central proton). Thus, electrons of all five orbitals contribute to the binding in the system. As a result, the bonds in $H_5O_2^+$ ion are energetically very close to those in HF_2^- ion.

Being highly sensitive to the arrangement of nuclei, the peculiar electron density distribution in these ionic structures survives only in the absence of external perturbation or when the perturbation has the same symmetry (no less than $\mathbb{C}_2$). However, for the spatial reasons, this kind of binding can by no means be spread on the residual structure in $(H_5O_2^+)(H_2O)_k$ and $(HF_2^-)(HF)_k$ systems, and all other hydrogen bonds are characterized by nearly the same electron density distribution as in neutral clusters.

Conclusion

Thus, hydrogen atom plays the role of a bridge between two neighboring electronegative atoms, i.e., acts as a binding joint between their electron density "clouds". This role is more pronounced in the ionic systems, but could already be deduced from the electron density distributions in neutral clusters. If we consider a general situation, the formation of an X–H... Y–Z hydrogen bond is determined by the overlap of X–H and Y–Z bonding orbitals (or Y lone pair orbital); and hydrogen atom acts as a bridge.

Typical hydrogen bonds are determined by both σ and π binding and can be classified as quasi-double and sometimes (as in the case of HF_2^- and $H_5O_2^+$ ions) quasi-triple.

H-bonds should be divided in two classes, namely, (i) separate bonds formed between two certain particles and (ii) bond networks that unite a large number of molecules and are stable when a delocalized conjugated π system appears. Networks can be formed by hydrogen fluoride and water molecules, as well as any molecules that involve sterically accessible O–H or N–H functional groups. The most stable building blocks of the networks are those with delocalized quasi-aromatic π systems.

The discovered peculiarities of the electron density distribution in H-bond domains clarify the origin of the NMR signatures of the bonds, particularly deshielding of H-bond protons and spin-spin interaction between X and Y. As we could see, electrons described by σ and π orbitals actually bind X and Y, especially when a conjugated system forms. At the same time, in many cases, zeros of both σ and π functions are close to the position of the bridge proton and accurately coincide with it in HF_2^- and $H_5O_2^+$ ions.

The existence of conjugated π system in molecular clusters determines the so-called cooperativity phenomenon. The spatial extension of cooperative effects depends on the extension of the delocalized conjugated π system, and the effects are most pronounced within the regions of quasi-aromatic π-binding typical of structural rings and closed or open hoops and strips, which consist of fused rings.

References

[1] Moore, T.S., Winmill, T.F., The state of amines in aqueous solution, *J. Chem. Soc. Trans.*, 101 (1912) 1635-1676.

[2] Huggins, M.L., Graduate Theses, University of California, 1919.

[3] Latimer, W.M., Rodebush, W.H., Polarity and ionization from the standpoint of the Lewis theory of valence, *J. Am. Chem. Soc.*, 42 (1920) 1419-1433.

[4] Pauling, L., *The Nature of the Chemical Bond and the Structure of Molecules and Crystals: an Introduction To Modern Structural Chemistry,* Cornell Univ. Press, New York, 1939.

[5] Bernal, J.D., Fowler, R.H., A theory of water and ionic solution, with particular reference to hydrogen and hydroxyl ions, *J. Chem. Phys.* 1 (1933) 515-548.

[6] Pauling, L., Corey, R.B., Branson, H.R., The structure of proteins: Two hydrogen-bonded helical configurations of the polypeptide chain, *Proc. Natl. Acad. Sci. USA,* 37 (1951) 205-211.

[7] Watson, J.D., Crick, H.H.C., *Molecular structure of nucleic acids: A structure for deoxyribose nucleic acid*, Nature, 171 (1953) 737-738.

[8] Arunan, E., Desiraju, G.R., Klein, R.A., Sadlej, J., Scheiner, S., Alkorta, I., Clary, D.C., Crabtree, R.H., Dannenberg, J.J., Hobza, P., Kjaergaard, H.G., Legon, A.C., Mennucci, B., Nesbitt, D.J., Definition of the hydrogen bond (IUPAC Recommendations 2011), *Pure Appl. Chem.*, 83 (2011) 1637-1641.

[9] Schmidt, M.W., Baldridge, K.K., Boatz, J.A., Elbert, S.T., Gordon, M.S., Jensen, J.H., Koseki, S., Matsunaga, N., Nguyen, K.A., Su, S., Windus, T.L., Dupuis, M., Montgomery, J.A.Jr., General atomic and molecular electronic structure system, *J. Comput. Chem.* 14 (1993) 1347-1363.

[10] Granovsky, A.A., Firefly version 7.1.G,
`http://classic.chem.msu.su/gran/firefly/index.html`.

In: Models in Bioscience and Materials Research ISBN: 978-1-62808-052-0
Editor: Kh. T. Kholmurodov

Chapter 9

DEVELOPMENT AND APPLICATION OF FRAGMENT-BASED ELECTRONIC STRUCTURE CALCULATION METHOD FOR BIOMOLECULE

Tomoki Kobori[1,*], Takao Otsuka[1], Shinji Tsuneyuki[2] and Makoto Taiji[1]
[1]Quantitative Biology Center (QBiC)
RIKEN (The Institute of Physical and Chemical Research)
Kobe, Hyogo, Japan
[2]Department of Physics, The University of Tokyo
Bunkyo-ku, Tokyo, Japan

Abstract

In this study we introduce an electronic structure calculation method for biomolecules, which we call FMO-LCMO method [1-2]. The term 'FMO' comes from Fragment Molecular Orbital [3-5]method, whose purpose is to calculate total energy and density matrix of the entire biomolecule with little computational cost. The reduction of CPU cost is accomplished by dividing the whole system into small subsystems called fragment.

The FMO-LCMO method was developed by combining the FMO method with the idea of Linear Combination of Molecular Orbitals

*E-mail address: kobori@riken.jp (Corresponding Author)

(LCMO). Just as Hamiltonian of any molecule is expressed with LCAO in conventional MO method, FMO-LCMO Hamiltonian is expressed with fragment molecular orbitals of each fragment. This property enables us to elucidate electronic structure of the whole system in terms of unit molecules such as amino acids or nucleotide. Then molecular orbitals and their orbital energies of the whole system can be calculated by diagonalization of the FMO-LCMO Hamiltonian. As our recent progress, we show how to calculate mulliken population with FMO-LCMO, and its calculation results for Chignolin with explicit water solvent. Charge distribution of both Chignolin and water molecules are also discussed.

Keywords: Electronic Structure, Calculation Method, Biomolecules, Molecular Orbitals

1. Introduction

A number of efforts have been made on simulation of biomolecules to study their structures and functions for years. Early studies mostly focused on the structures to simulate behavior of biomolecules based on classical molecular mechanics (MM). Although MM method has been used successfully for a wide range of huge molecules, the method immediately confronts with a serious problem that electronic density change such as charge transfer and chemical reactions can not be considered in this method. Quantum mechanical (QM) methods have been developed and used to overcome such problems, while the most difficult part in that method is increasing computational cost that scales as $O(N^3) \sim O(N^4)$ with N being the number of electrons in the system.

A hybrid method that combines the above two methods is known as QM/MM method. Since interesting part in a molecule is often limited such as substrate and binding pocket of an enzyme, QM/MM method describes such important part with QM method, and remainder with MM method. However, one problem in QM/MM method is to inherit the difficulty of MM method. The environment is still dealt with by MM method, and it is expected that the contribution from the environment becomes larger as the system size grows.

In order to calculate the whole system with quantum mechanics, so-called Order-N methods have been proposed. In these methods, linear scaling algorithm is introduced by dividing a molecule anyway for the purpose of restricting its amount of computation to $O(N)$. Although arbitrariness remains with respect to where to divide the molecule, the merit of this method is that the whole system can be calculated by quantum mechanical method.

Generally speaking, Order-N method firstly divides a molecule into fragments, then calculations are carried out on each fragment, and finally the desired properties of the whole system are evaluated with those of the fragments. Although above feature is shared in many types of Order-N methods as a whole, there are many kinds of versions in practice.

The divide-and-conquer (DC) method proposed by W. Yang also divides the system into subsystems, and then calculations on all the subsystems are carried out using a local Hamiltonian. The density matrix of the whole system is constructed by combining the density matrices of the subsystems. Then total energy and other properties of the system can be calculated as the expectation value of the Hamiltonian operator. The key in the DC method is that subsystem calculations are performed with its surrounding spaces called buffer region, and electron chemical potential of the whole system is determined so that number of electrons conserves over the whole system.

Fragment molecular orbital (FMO) method, which was proposed by Kitaura *et al*, can also be classified into such Order-N methods. Here we explain about only remarkable hallmarks of the FMO method compared to other Order-N methods. In the FMO method, calculations of all the fragments are performed under the condition that these fragments are immersed in the electronic field generated by the environment. Thus the many-body effects of the environment can be considered. Moreover, all the fragment pairs and optionally trimers are also calculated to achieve desirable accuracy. The way of fragmentation does not need any additional atoms to terminate dangling bonds. Instead, an sp^3 carbon atom is divided into two atoms and its electrons and protons are reassigned to each fragment.

Despite lots of approximate treatments in the theory, the FMO method gives rather accurate total energy compared to conventional *ab initio* calculation, as far as the system is non-metallic like biomolecules. It is said that required accuracy in protein science is about 1 kcal/mol. The FMO method can satisfy this accuracy condition.

However, FMO itself can not calculate molecular orbitals and their orbital enegies. This is because FMO does not construct Hamiltonian of the whole system. Therefore we developed a method which can construct Hamiltonian of the whole system based on the FMO method. We named this method FMO-LCMO method, since Hamiltonian of the whole system is expressed with linear combination of molecular orbitals (LCMO) of each fragment. In this chapter, we especially show FMO-LCMO-based mulliken population analysis. From here we explain FMO-LCMO method, how to construct Hamiltonian of the whole system and how to calculate fragment-based mulliken population.

FMO method basically evaluates total energy of the whole system following the total energy expansion. In the case of FMO2, this expansion form can be transformed into

$$E^{\mathrm{FMO2}} = \sum_{I>J} E_{IJ} - (N_f - 2) \sum_{I} E_I. \tag{1}$$

This means total energy of the whole system can be obtained by the sum of dimer energies from which the double-counting parts are subtracted by using monomer energies. Similarly, total energy in FMO3 can be changed into the following expression.

$$E^{\mathrm{FMO3}} = \sum_{I>J>K} E_{IJK} - (N_f - 3) \sum_{I>J} E_{IJ} + \frac{1}{2}(N_f - 2)(N_f - 3) \sum_{I} E_I. \tag{2}$$

The above expression Eqs.(1) and (2) are devised to cancel the double counting of the environmental potential, as well as ordinary interfragment interaction.

Although FMO incorporates a lot of approximate treatments such as omitting SCF calculation of far separated n-mer ($n = 2, 3$), and neglect of exchange-correlation term in the environmental potential, it is well known that results calculated by FMO agree very well with those of conventional one-electron theory.

However, FMO itself does not have an explicit definition of wave functions and orbital energies of the whole system. Therefore one remaining problem in FMO is to establish the way of calculating MO and MO energies of the whole system. In this respect, FMO-MO was proposed as a new method of calculating MOs based on FMO, and it is reported that one-electron orbital energies and their wave functions calculated by FMO-MO are in good agreement with those obtained by conventional method.

Using density matrix of the whole system calculated by FMO method, FMO-MO constitutes Hamiltonian of the whole system through only once SCF iteration, while one difficulty is that FMO-MO method has to carry out all the two-electron integrals over the whole system.

On the other hand, the FMO-LCMO method we propose here enables us to construct one-electron Hamiltonian of the whole system by directly assembling subsystem Hamiltonians. Since the construction of the total system Hamiltonian in FMO-LCMO is performed without calculating two-electron integrals, it requires far less computational cost than FMO-MO. Furthermore, FMO-LCMO can drastically reduce the cost of diagonalization of Hamiltonian matrix by focusing on HOMO/LUMO levels. It also gives us an intuitive understanding of electronic structure of molecules since molecular orbitals of the whole system are realized by linear combination of molecular orbitals (LCMO) of each fragment.

2. Results and Discussion

The basic framework of the FMO-LCMO method is explained in this chapter. Here we introduce simple notation of FMOn-LCMO($n = 2, 3$). FMO-LCMO without fragment trimer effect is denoted by FMO2-LCMO, and that including trimer by FMO3-LCMO.

In the FMO-LCMO method, Hamiltonian of the whole system is defined by combination of each fragment and interfragment Hamiltonian submatrices as Fig. 2 shows.

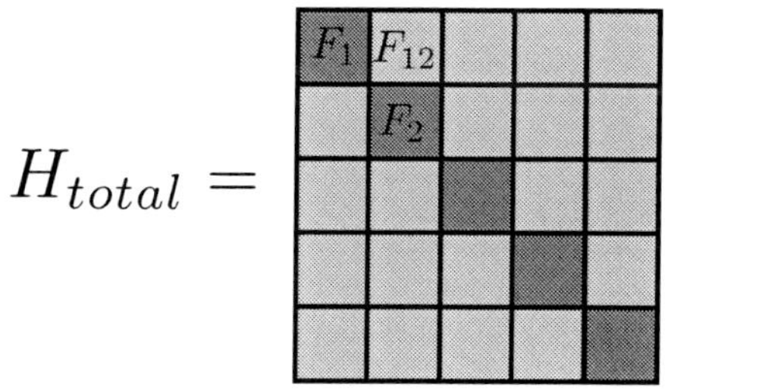

Figure 1. Conception of Hamiltonian of the whole system in FMO-LCMO method. Diagonal block (red block) is fragment Hamiltonian, Off-diagonal block (blue block) is interfragment Hamiltonian.

Furthermore, Hamiltonian matrices of each fragment and interfragment are defined as

$$(F_I)_{pq} = \sum_{J \neq I} (H_{I \leftarrow IJ})_{pq} - (N_f - 2)(H_I)_{pq}, \tag{3}$$

where F_I is a fragment Hamiltonian matrix with which I-th diagonal block of the whole system is finally covered. It should be mentioned that the effect of all dimers including I-th monomer can be fully incorporated into this submatrix. By this procedure, we can fully utilize all the fragment Hamiltonians calculated by the FMO.

Similarly, for off-diagonal blocks ($I \neq J$), following expression can be adopted.

$$(F_{IJ})_{pq} = (H_{IJ})_{pq}. \tag{4}$$

If we consider three-body term which is derived from fragment-trimer calculation, each fragment and interfragment Hamiltonian is given by the following expression instead of the above expression.

Diagonal block:

$$\begin{aligned}(F_I)_{pq} &= \sum_{\substack{J<K \\ J,K\neq I}}\sum (H_{I\leftarrow IJK})_{pq} \\ &- (N_f - 3)\sum_{J\neq I}(H_{I\leftarrow IJ})_{pq} + \frac{1}{2}(N_f - 2)(N_f - 3)(H_I)_{pq},\end{aligned} \tag{5}$$

The off-diagonal regions ($I \neq J$), interfragment Hamiltonian matrices are

$$(F_{IJ})_{pq} = \sum_{K\neq I,J} (H_{IJ\leftarrow IJK})_{pq} - (N_f - 3)(H_{IJ})_{pq} \tag{6}$$

It should be noted that indices p, q do not mean atomic orbital, but molecular orbital of each fragment.

After Hamiltonian of the whole system is constructed using $\{F_I\}$ and $\{F_{IJ}\}$, molecular orbitals and their orbital energies of the whole system can be obtained by diagonalization of FMO-LCMO Hamiltonian H_{total} and its corresponding overlap matrix S_{total}. Since molecular orbital coefficients can be calculated by diagonalizing FMO-LCMO Hamiltonian, fragment-based mulliken population can also be calculated using the MO coefficients. Paying attention to the fact that basis function p, q does not denote atomic orbitals (AO) but fragment molecular orbitails (FMO), number of electrons of the whole system N can be divided into number of electrons which belong to I-th fragment N_I as follows.

$$
\begin{aligned}
N &= 2\sum_{i}^{occ.}\int|\psi_i|^2 d\mathbf{r} \\
&= 2\sum_{i}^{occ.}\sum_{p,q} C_{pi}^{*}C_{qi}\langle\phi_p|\phi_q\rangle \\
&= \sum_{pq} D_{pq}S_{qp} = \sum(\mathbf{DS})_{pp} \\
&= \sum_{I}\sum_{p\in I}(\mathbf{DS})_{pp} = \sum_{I} N_I.
\end{aligned} \tag{7}
$$

where molecular orbital of the whole system is expanded with FMO basis functions like $\psi_i = \sum\limits_{p} C_{pi}\phi_p$, and density matrix $\mathbf{D}$ and overlap matrix $\mathbf{S}$ are also defined as $D_{pq} = \sum\limits_{i}^{occ.} C_{pi}^{*}C_{qi}$, $S_{pq} = \langle\phi_p|\phi_q\rangle$, respectively.

From here we demonstrate application example of FMO-LCMO method and its mulliken charge for Chignolin molecule surrounded by spherically distributed water molecules. Chignolin is an artificial protein composed of 10 amino acids (GYDPETGTWG), and it is known that this protein forms a particular stable β-hairpin structure in water. Here we adopted FMO2-LCMO method, B3LYP functional and 6-31G* basis set is used. Fig. 2 shows molecular model which we calculated in this study.

In Fig. 2, any of these three models are constructed by extracting particular water molecules inside a sphere whose radius is r(r =10 Å, 15 Å, 20 Å) from the previously-prepared system, which consists of Chignolin molecule and enough number of surrounding water molecule generated by TIP3P. After generating water molecules by TIP3P around Chignolin, energy-minimization is performed for only water molecules then Chignolin and only waters inside the sphere is extracted to make the model.

Fig. 2 is mulliken charge distribution for water molecules. The 44 water molecules in Fig. 2 are those which are common for $r = 10, 15, 20$ Å. In other words, these 44 water molecules are located inside the smallest sphere whose radius is $r = 10$ Å. As this figure shows, 21-th and 28-th water molecule have a large charge in the case of $r = 10$ Å. This is because in the case of $r = 10$ Å, there are no water molecules outside these two water molecules. When it comes to $r = 10, 15$ Å, these two water molecules do not have large charge any more.

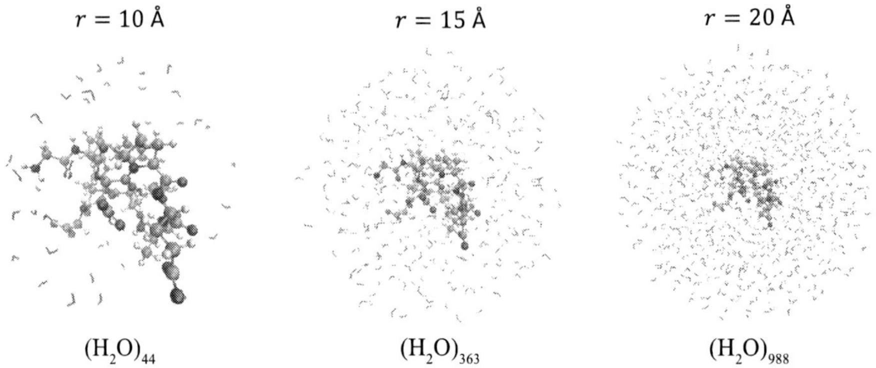

Figure 2. Model system of Chignolin molecule and surrounding water molecules.

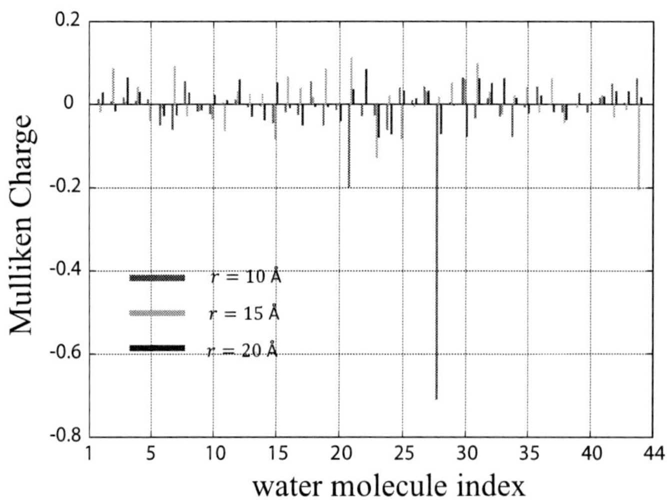

Figure 3.

This is because outer side of water molecules extinguish charge polarization of the two water molecules. In the case of $r = 10, 15$ Å, there are no peaks whose absolute value of mulliken charge is much larger than 0.2. What is a little surprising is that distribution of $r = 15$ Åand $r = 20$ Ådoes not agree with each other so much. This means water molecules located in the 15 Å$< r < 20$ Åspherical shell have an electronic influence on those located in the 10 Å$< r <$

15 Å. However, considering absolute values of mulliken charge for almost all of the water molecules in the case of $r = 15, 20$ Åare less than 0.1, the difference between $r = 15$ Åand $r = 20$ Åmay not be worth investigating.

Fig. 2 shows mulliken charge of each fragment of Chignolin in vacuum or water sphere. Just like Fig. 2, regarding solvent effect, we demonstrate three cases of $r = 10, 15, 20$ Å. Firstly, one outstanding feature is mulliken charge of 4-th fragment (Glu), the value is 1.97. Another feature is that mulliken charge distributions of vacuum, $r = 15, 20$ Åare almost same as each other. From this results, we can conclude that $r = 10$ Åis not enough to reproduce solvent effect.

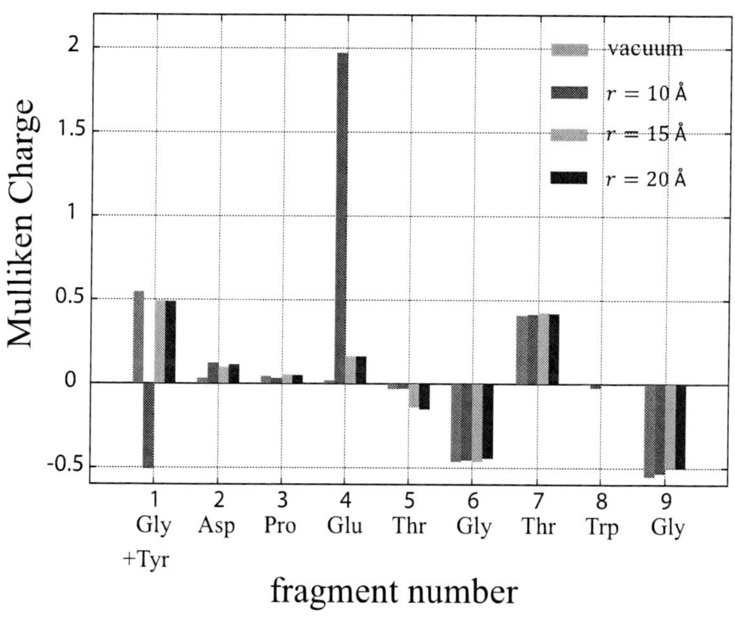

Figure 4.

Acknowledgments

I am deeply grateful to Dr. Kholmirzo Kholmurodov, who kindly invited me for MSSMBS'12 and provided us with a great suggestion.

References

[1] S. Tsuneyuki, T. Kobori, K. Akagi, K. Sodeyama, K. Terakura, "Molecular orbital calculation of biomolecules with fragment molecular orbitals", *Chem. Phys. Lett.*, 476 (2009) 104.

[2] H. Nishioka, K. Ando, "Electronic coupling calculation and pathway analysis of electron transfer reaction using *ab initio* fragment-based method. I. FMO-LCMO approach", *J. Chem. Phys.*, 134 (2011), 204109

[3] K. Kitaura, E. Ikeo, T. Asada, T. Nakano, M. Uebayasi, "Fragment molecular orbital method: an approximate computational method for large molecules", *Chem. Phys. Lett.*, 313 (1999) 701-706.

[4] D. G. Fedorov, K. Kitaura, "The importance of three-body terms in the fragment molecular orbital method", *J. Chem. Phys.*, 120 (2004) 6832-6840.

[5] T. Nakano, T. Kaminuma, T. Sato, Y. Akiyama, M. Uebayasi, K. Kitaura, "Fragment molecular orbital method: application to polypeptides", *Chem. Phys. Lett.*, 318 (2000) 614-618.

In: Models in Bioscience and Materials Research ISBN: 978-1-62808-052-0
Editor: Kh. T. Kholmurodov

Chapter 10

MOLECULAR DYNAMICS SIMULATIONS FOR SMALL-ANGLE NEUTRON SCATTERING: SCATTERING LENGTH DENSITY SPATIAL DISTRIBUTIONS FOR MONO-CARBOXYLIC ACIDS IN D-DECALIN

Roman A. Eremin*[1,*]*, Kholmirzo T. Kholmurodov*[1,2]*,
***Viktor I. Petrenko*[1,3]*, László Rosta*[4] *and Mikhail V. Avdeev*[1]**
[1]Joint Institute for Nuclear Research, Dubna, Moscow Region, Russia
[2]International University Dubna, Dubna, Moscow Region, Russia
[3]Taras Shevchenko National University, Kyiv, Ukraine
[4]Wigner Research Centre for Physics, Budapest, Hungary

Abstract

The method of the neutron scattering length density (SLD) spatial distributions calculations based on the molecular dynamics simulation data is described on the example of saturated mono-carboxylic acids in decalin solutions. The sufficient SLD modulation in the solute-solvent interface area is obtained. Results of the current work and described calculation method can be used for small-angle neutron scattering data analysis.

Keywords: Small-Angle Neutron Scattering, Molecular Dynamics, Mono-Carboxylic Acids, d-Decalin

*E-mail address: era@jinr.ru (Corresponding Author)

1. Introduction

Small-angle neutron scattering (SANS) is an important method for condensed matter investigation due to specific properties of neutrons. Characteristic size of the inhomogeneities of the scattering length density (SLD) which can be studied with SANS is from 1 to 100 nm. This relatively wide size range includes quietly different systems: biological, chemical, physical, etc. SANS is based on the incident thermal neutron beam widening due to coherent scattering processes and the analysis of the scattered neutron flux. It is worth noting that modeling and interpretation of the data obtained with SANS (differential cross-section per sample volume, or scattering intensity I, as a function of the momentum transfer module q) are often complicated by incoherent scattering, e.g., by hydrogen in the systems. Furthermore, in case of solutions if the scattering particles size is close to the method resolution lower limit, there is no possibility to neglect solvent reorganization at the solute-solvent interface area (in comparison with pure state), hence the field mentioned above influence on the scattering is possible.

Classical molecular dynamics simulation (MDS) provides information about position of each atom in the systems under studying within an interesting period of time. For this reason, in case of the Van der Waals and Coulomb interactions MDS can be used for solute-solvent area investigation in terms of SLD spatial distribution. The aim of the current review is to introduce the readers with the method of SLD spatial profiles calculations and subsequent using in case of anisotropic scattering particles. The elements of the self-made *Mathematica* code concerning calculations mentioned above are represented below and, authors believe, can be easily ported to other languages. According to the purpose the saturated mono-carboxylic acids non-polar (d-decalin based) dilute solutions were chosen because the solute-solvent interface area influence on the SANS has been indirectly obtained thought the Guinier approximation (GA) application for these systems early. For this reason MDS-based SANS data modeling could potentially be used within the analysis instead the GA.

2. Experimental and Simulation Methods

2.1. SANS

Pure myristic (MA) and stearic (SA) (both 99.9%) acids (Merck) were dissolved in d-decalin, $C_{10}D_{18}$ (99.8%, Deutero GmbH) with ultrasonication. SANS measurements (at 25°C) were done for 3% (MA) and 2% (SA) solutions regarding the volume fraction of the solutes, Φ. These concentrations were chosen, on the one hand, to diminish the structure-factor effect on the scattering curves as a consequence of the acid interaction in the solutions, and, on the other hand, to exceed the sensitivity limit of the SANS method and get statistically reliable data. The SANS experiment was performed on the Yellow Submarine small-angle instrument at the steady-state reactor of the Budapest Neutron Centre (BNC), Hungary. The scattered intensity was obtained as a function of the momentum transfer module, $q = (4\pi/\lambda)sin(\theta/2)$, where λ is the incident neutron wavelength, and θ is the scattering angle. The fixed wavelength of 0.386 nm (monochromatization $\Delta\lambda/\lambda = 20\%$) and sample-detector distances of 1.3 and 5.6 m (detector size 0.64 m) were used to cover the q-interval of $0.15 - 4$ nm^{-1}. The calibration with a 1-mm water sample was made in a standard way[1] after the corrections for the background, buffer (d-decalin) and empty quartz sample cell.

In Fig.1 the obtained SANS curves are compared with previously measured for d-benzene ones and GA approximation Eq.1 which consists in addiction to exponential member the bkg parameter (incoherent background value).

$$I(q) \approx I(0)exp(-q^2R_g^2/3) + bkg. \tag{1}$$

Fig.1 also shows curves of Eq.1 fitting best the experimental ones when the Guinier parameters are varied together with bkg (the final parameter values are indicated in the caption). In [2] the correctness of such treatment for benzene-based solutions was testified by two factors. Firstly, the experimentally obtained dependence of the bkg-value on the volume fraction of the dissolved acid molecules was strongly linear; secondly, the scattering curves referred to one acid concentration coincided at sufficiently large momentum transfer values. However, when we compare the bkg-values obtained in this way (curves constant values in the $q \geq 4$ nm^{-1}) for the solutions under study, there are significant differences between them. At the same time, the bkg parameter should be at one level for the same acid with the same concentration in the two solvents, since for the considered curves (after the subtraction of the solvent scattering) it is determined only by the hydrogen content in the acid molecules. The found differences are indicative of some systematic errors in the determination of the background. Especially it concerns the d-decalin solutions, where bkg-values are above those for the d-benzene solutions. There are several factors with possible effects on the SANS modeling. One of them can be related to the fact

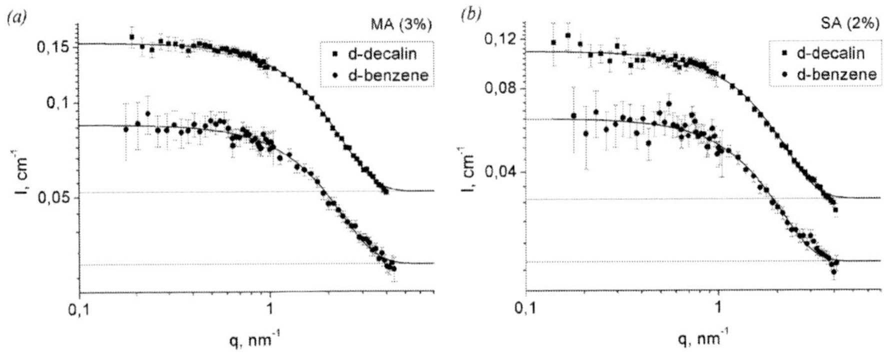

Figure 1. Experimental SANS curves for myristic (a) and stearic (b) acids in d-benzene and d-decalin; best fits according to Eq. 1 are shown by solid lines. The results of the fits (parameter errors in the last significant figure are indicated in parentheses) are MA: $I(0) = 0.054(1)$ cm^{-1}, $R_g = 0.88(2)$ nm, $bkg = 0.031(1)$ cm^{-1} (benzene); $I(0) = 0.103(1)$ cm^{-1}, $R_g = 0.89(1)$ nm, $bkg = 0.052(1)$cm^{-1} (decalin), SA: $I(0) = 0.043(1)$ cm^{-1}, $R_g = 1.00(2)$ nm, $bkg = 0.019(1)$cm^{-1} (benzene); $I(0) = 0.077(1)$ cm^{-1}, $R_g = 0.97(1)$ nm, $bkg = 0.032(1)$ cm^{-1} (decalin). The bkg-values represented by thin lines.

that decalin molecules are comparatively large and anisotropic (as compared to benzene ones) hence the interface organization of the solvent molecules around a thin and rod-like solute can contribute significantly to the scattering.
For the reason mentioned above the studying of the interface area in terms of the SLD spatial distribution for decalin based solutions is very important in order to perform correct SANS data interpretation. Sure enough MDS can shed the light on solute-solvent interface are effect on the scattering in such formulation of the problem.

2.2. MDS

In Fig. 2 the single d-decalin molecule (a) and the whole system under investigation (myristic acid in d-decalin) (b) are represented at the same scale.

MDS on the limiting (one solute molecule per MD cell) solutions of MA and SA was performed using the DL_POLY 2.20 general-purpose code [3] at the parallel cluster of Central Information and Computer Complex of the Laboratory of Information Technologies, Joint Institute for Nuclear Research (CICC

(a)

(b)

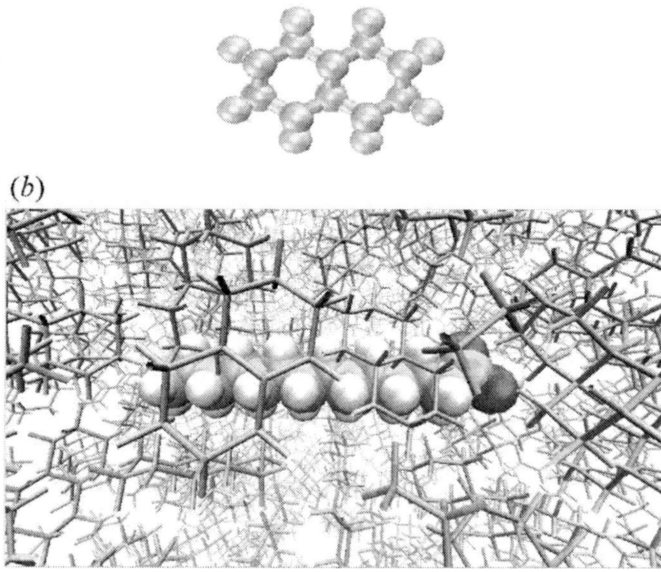

Figure 2. Snapshots of single d-decalin molecule (a) and the whole system (b) under investigation on example of myristic acid in d-decalin.

LIT JINR). The intermolecular interaction was modeled as a sum of van der Waals (described by two-parameter Lennard-Jones (LJ) potential) and electrostatic interactions [4] with cutoff radii equal to 1 and 2 nm, respectively:

$$U = \sum_{i,j \neq i} 4\epsilon_{ij} \left(\left(\frac{\sigma_{ij}}{r_{ij}} \right)^{12} - \left(\frac{\sigma_{ij}}{r_{ij}} \right)^{6} \right) + \frac{Q_i Q_j}{r_{ij}}, \tag{2}$$

where σ_{ij} and ϵ_{ij} are the size and energy parameters of the LJ interaction between i^{th} and j^{th} atoms positioned at the distance r_{ij} between them; Q_i and Q_j are their partial charges. The parameters of the interaction between non-identical atoms were found according to the standard Lorentz-Berthelot combination rule:

$$\sigma_{ij} = \frac{\sigma_i + \sigma_j}{2}; \epsilon_{ij} = \sqrt{\epsilon_i \epsilon_j} \tag{3}$$

For the atoms in the acid molecules the following LJ potential parameters were used [5]: $\sigma_C = 0.392$ nm, $\epsilon_C = 0.12$ kcal/mol, $\sigma_H = 0.242$ nm, $\epsilon_H = 0.02$ kcal/mol, $\sigma_O = 0.281$ nm, $\epsilon_O = 0.20$ kcal/mol. The LJ parameters for d-decalin (regardless of cis or $trans$ stereoisomers) were found previously by MDS of the pure solvent [6]: $\sigma_C = 0.335$ nm, $\epsilon_C = 0.06$ kcal/mol, $\sigma_D = 0.225$ nm, $\epsilon_D = 0.02$ kcal/mol.

The electric charge distributions in the molecules of deuterated $cis/trans$-decalin and acids were determined using a semi-empirical quantum chemistry method AM1 (Austin Model 1). The resulting average values of the partial charges in the electron charge value are $Q_D = 0.08$, $Q_C = -0.14$ for decalin and $Q_H = 0.10$, $Q_C = -0.15$, $Q_O = -0.34$ for acids.

For pure deuterated decalin system the crystal structure corresponding to protonated $trans$-decalin [7] was used as an initial state. In order to model the mixture of isomers, half of molecules were replaced by cis isomers. The simulation cell (parallelepiped $5.5 \times 5.3 \times 5.3$ nm^3) contained 700 decalin molecules. To model the limiting acid solutions, one molecule of MA or SA was placed in the centre of the cell and all decalin molecules overlapped with it were removed. The molecules of decalin and the acids were considered as rigid units with constant bond lengths.

The modeled systems (pure decalin and solutions) were relaxed during 100 ps in the Hoover NPT ensemble (particle number, pressure (atmospheric) and temperature (300 K) all being constant); the relaxation time constant of barostat and thermostat was 0.3 ps. After the relaxation under normal conditions the atomic coordinates were recorded every 1 ps in the 2 ns trajectory during the modeling in the Nose-Hoover NVT ensemble (constant particle number, volume and temperature) in order to perform subsequent SLD distribution calculations. The Ewald algorithm was applied to treat the long-range electrostatic interactions. The Verle integration algorithm for the numerical solution of the equations of motion was realized with the time step of 0.001 ps for both ensembles.

3. Spatial SLD Calculation from Trajectory File

Now we start the description of the spatial SLD distributions calculation based on the data found in the trajectory file (HISTORY in case of DL_POLY). As it was mentioned 1000 steps (1 ps) was chosen as a period of the atomic coordinates putting into the trajectory file. Atomic density distribution determines the SLD one uniquely. There are few code fragments (Listing 1-4) with corresponding descriptions below.

First of all you should set the MDS initial parameters (requested duration of the simulation as a number of time steps, equilibration time interval and trajectory file interval) manually or in order to make the code as general as possible read them automatically (e.g. from the OUTPUT data file). The code fragment of such reading is presented in Listing 1.

Listing1

```
ClearAll["Global'*"]
SetDirectory["Decalin/AM1/D/acids/NVT1/ma"];
str = OpenRead["OUTPUT"];
Find[str, "number of timesteps"];
StringSplit[SetStreamPosition[str, StreamPosition[str] -
- StringLength[Last[NTS = Read[str, Number];
Find[str, "equilibration period"];
StringSplit[SetStreamPosition[str, StreamPosition[str] -
- StringLength[Last[ EP = Read[str, Number];
Find[str, "trajectory file interval" ];
StringSplit[ SetStreamPosition[str, StreamPosition[str] -
- StringLength[Last[ TFI = Read[str, Number];
Close[str];

str = OpenRead[ "HISTORY" ];
c1 = 3; c2 = 4; c3 = 8; c4 = 9;
step = 1/2;
Xmax = 40; Ymax = 30; Zmax = 30;
atompos = {};
sorts = {"D", "C", "H", "O"};
scatl = {6.671, 6.646, -3.739, 5.803};
res = Array[0 &, {Xmax/step, Ymax/step, Zmax/step} ];
For[i = 1, i <= Nshell, i++, {
For[j = 1, j <= Length[sorts], j++, ress[i, j] = 0 ]
} ]
conf = 0;
Print[Dynamic[Round[100conf/((NTS - EP)/TFI + 1) ]],
"Print[Dynamic[conf], " th configuration is in process" ]
Print[Natoms, " ", Dynamic[i]];
```

Then you should open the required HISTORY file to read and set the number of atoms which are consequently used for coordinate systems (CS) initialization. Simple Cartesian CS is used. Acid molecules were represented as rigid units within the simulation. Due to this simplication it is convenient to orientate one of the Cartesian CS axes among the acid alkyl chain (e.g. X axis). In general at least three quite distant from each other atoms could be chosen for this purpose. Let us no loss the generality to consider MA solutions in decalin as an example.

In Listing 1 you can see the four number of carbon atoms in the MA alkyl chain that were chosen to calculate the required rotational angles. Then the step of the spatial SLD distribution calculations and the sizes of the field around the MA molecules are specified in angstroms (internal DL_POLY units of the lengths) among each of the Cartesian CS axes. Two arrays (denoted by $sorts$ and $scatl$) consist of elements types and corresponding scattering length values, respectively. There is initialization of the res (array of the result). Print functions are not necessary but useful for the program evaluation monitoring.

The main program starts now (see Listing 2). There is the For-loop by configurations available in trajectory file. At the first step the program reads the MD cell size (lx, ly, lz) and $trans$ array is created (translation array). The meaning of the last one will be described below. After that position of the first carbon atom from the set is chosen as a real acid position. The positions of the rest of the carbon atoms set should be calculated taking into account possible acid crossing of the MD cell border. In the other words we must apply periodic boundary conditions. For this reason position of the other carbon atoms are corrected taking into account real distances to the first one available from the acid molecule structure (invariable within the simulation). Now we can perform calculations of the rotational angles $\alpha z, \alpha y, \alpha x$ corresponding to rotation around z, y and x axis in order to provide desired acid molecule position. When three angles are found program returns to the configuration start by $SetStreamPosition$ command.

Listing2

```
For[ts = 1, ts ≤ NTS, ts += TFI, {
str1 = Find[str, "timestep" ];
conf++;
Natoms = ToExpression[StringSplit[str1][[3]]];
lx = ToExpression[StringSplit[Read[str, String]][[1]]];
ly = ToExpression[StringSplit[Read[str, String]][[2]]];
```

```
lz = ToExpression[StringSplit[Read[str, String]][[3]]];
strpos = StreamPosition[str];
trans = {};
For[tx = 1, tx ≥ -1, tx--, {
For[ty = 1, ty ≥ -1, ty--, {
For[tz = 1, tz ≥ -1, tz--, {
AppendTo[trans, {tx lx, ty ly, tz lz} ]} ]} ]} ];
trans = Join[Take[Sort[trans], -14 ]; Take[Sort[trans], 13
]];
For[i = 1, i ≤ c1, i++, Find[str, "CM " ]];
s = Read[str, String]; s = StringReplace[s, "E" →
"*10^" ];
r4 = ToExpression[StringSplit[s]];

For[i = c1 + 1, i ≤ c2, i++, Find[str, "CM " ]]; s =
Read[str, String]; s = StringReplace[s, "E" → "*10^" ];
r2 = ToExpression[StringSplit[s]]; For[tr = 1, tr ≤ 27,
tr++, { r2cor = r2 + trans[[tr]];
If[Abs[EuclideanDistance[r4, r2cor] - 1.51 ] < 10^-1, r2 =
r2cor]; } ];

For[i = c2 + 1, i ≤ c3, i++, Find[str, "CM " ]]; s =
Read[str, String]; s = StringReplace[s, "E" → "*10^" ];
r1 = ToExpression[StringSplit[s]]; For[tr = 1, tr ≤ 27,
tr++, { r1cor = r1 + trans[[tr]];
If[Abs[EuclideanDistance[r4, r1cor] - 6.31 ] < 10^-1, r1 =
r1cor]; } ];

For[i = c3 + 1, i ≤ c4, i++, Find[str, "CM " ]]; s =
Read[str, String]; s = StringReplace[s, "E" → "*10^" ];
r3 = ToExpression[StringSplit[s]]; For[tr = 1, tr≤ 27,
tr++, { r3cor = r3 + trans[[tr]];
If[Abs[EuclideanDistance[r4, r3cor] - 7.50 ] < 10^-1, r3 =
r3cor]; } ];

centr = Mean[{r1, r2, r3, r4} ];
r1 -= centr; r2 -= centr; r3 -= centr; r4 -= centr;


max = 0; αz = 0;
For[i = 0, i ≤ 2π, i += 0.01,
{If[ (RotationMatrix[i, {0, 0, 1}].(r2 + r4)) [[1]] > max,
{max = (RotationMatrix[i, {0, 0, 1} ].(r2 + r4)) [[1]],
αz = i} ]}];
r1 = RotationMatrix[αz, {0, 0, 1}].r1;
r2 = RotationMatrix[αz, {0, 0, 1}].r2;
r3 = RotationMatrix[αz, {0, 0, 1}].r3;
r4 = RotationMatrix[αz, {0, 0, 1}].r4;
```

```
max = 0; αy = 0;
For[i = −π/2, i ≤ π/2, i += 0.01,
{If[ (RotationMatrix[i, {0, 1, 0}].(r2 + r4)) [[1]] > max,
{max = (RotationMatrix[i, {0, 1, 0} ].(r2 + r4)) [[1]],
αy = i} ]}];
r1 = RotationMatrix[αy, {0, 1, 0}].r1;
r2 = RotationMatrix[αy, {0, 1, 0}].r2;
r3 = RotationMatrix[αy, {0, 1, 0}].r3;
r4 = RotationMatrix[αy, {0, 1, 0}].r4;

max = 0; αx = 0;
For[i = 0, i ≤ 2π, i += 0.01,
{If[ (RotationMatrix[i, {1, 0, 0}].r1) [[2]] > max,
{max = (RotationMatrix[i, {1, 0, 0} ].r1) [[2]], αx = i}
]}];
r1 = RotationMatrix[αx, {1, 0, 0}].r1;
r2 = RotationMatrix[αx, {1, 0, 0}].r2;
r3 = RotationMatrix[αx, {1, 0, 0}].r3;
r4 = RotationMatrix[αx, {1, 0, 0}].r4;


For[i = 1, i ≤ Natoms, i++, {
s = Read[str, String];
sr = Read[str, String];
sr = StringReplace[s, "E" → "*10^" ];
rr = ToExpression[StringSplit[sr]];
flag = 1;
For[tr = 1, tr ≤ 27, tr++,
If[EuclideanDistance[rr + trans[[tr]], centr] ≤
Norm[{Xmax/2, Ymax/2, (Zmax/2)} ],
{flag = 0; r = rr + trans[[tr]];} ]];

If[flag == 1, Continue[]];
r -= centr;

r = RotationMatrix[αx, {1, 0, 0} ].RotationMatrix[αy,
{0, 1, 0} ]
.RotationMatrix[αz, {0, 0, 1}].r;
If[(Abs[r[[1]]]              ≤ Xmax/2) && (Abs[r[[2]]] ≤ Ymax/2)
&& (Abs[r[[3]]] ≤ Zmax/2), {p = Position[sorts,
First[Characters[s]]][[1]][[1]];
iii = IntegerPart[ (r[[1]] + Xmax/2)/step ] + 1;
jjj = IntegerPart[ (r[[1]] + Ymax/2)/step] + 1;
```

```
kkk = IntegerPart[ (r[[1]] + Zmax/2)/step] + 1;
res[[iii]][[jjj]][[kkk]] += scatl[[p]]10^-13/N[step^3 10^-24 10^10]}];}
]; } ]; Close[str]; res /= conf;
```

There is nested loop with counter used for reading atom specification and position one-by-one and performing corresponding analysis. First of all, through the *Boolean*-like *flag* we can screen out atoms are not placed in the vicinity of acid molecules using only one *If* operation and taking into account all possible transitions from *trans*. If the atom under evaluation is close to acid molecule its position will be processed (acid center position subtraction, rotation to new CS, additional checking with three *If* operations) and the SLD value caused by this atom presence will be added to corresponding cell of result array. The last one is uniquely related with the geometry of the space field under consideration. After reading of all available configurations (*conf* is the counter) *res* must be normalized.

In order to minimize time consuming you can save *res* or the function (*func*) corresponding to shifted *res*. An example of such kind saving is presented in Listing 3.

Listing3

```
func = {};
For[i = 1, i ≤ Xmax/step, i++, {
For[j = 1, j ≤ Ymax/step, j++, {
For[k = 1, k ≤ Zmax/step, k++, {

AppendTo[func, {{(i - 1/2) step - Xmax/2, (j -
1/2) step - Ymax/2, (k - 1/2) step - (Zmax/2)},
res[[i]][[j]][[k]]}}]; }] }] }]
str = OpenWrite[ "FUNC" ]; Write[str, func]; Close[str]
```

Consequent using and the ways of the visualization of the obtained spatial SLD distributions are presented in the next subsection.

4. SLD Spatial Distributions Profiles

It is worth noting that selection of the *step* parameter value depends on the type of the modeling systems and the number of configurations available in the trajectory file. The last one in its turn is related strong with the characteristic period of neutron interaction with the system.

Listing4

```
str = OpenRead[ "Decalin/AM1/D/acids/FUNC-ma"];
func = Read[str]; Close[str]; Xmax = 40; Ymax = 30; Zmax
= 30;
scatl = {6.671, 6.646, -3.739, 5.803};
volumes = 185000;(*pure solvent volume in relaxed
state*)
mols = 700; (*solvent molecule number in the pure
solvent MD cell*)
ρs= 10^-10 (18 scatl[[1]] + 10 scatl[[2]]) 10^-13 (*fm→cm*) *mols/(volumes
10^-24
(*A^3 → cm^3*)); data = Table[{func[[60 60 (i - 1) + 60 (j
- 1) + k]][[1]][[1]],
func[[60 60 (i - 1) + 60 (j - 1) + k]][[1]][[2]],
func[[60 60 (i - 1) + 60 (j - 1) + k]][[1]][[3]],
func[[60 60 (i - 1) + 60 (j - 1) + k]][[2]]}, {i, 1,
Xmax/step},{j, 1, Ymax/step},{k, 1, Zmax/step}];
f = BSplineFunction[data, SplineDegree→5]


contset = {-7, -4, 0.2, 1, 2, 3, 5};
contshad = Table[Hue[0, 0, 0, i], {i, -1/4, 1, 1/(2
Length[contset]-4)}];
ContourPlot[f[(10x)/Xmax + 0.5, (10y)/Ymax + 0.5,
0.5][[4]]      -      ρs, {x, -(Xmax/20), (Xmax/20)}, {y,
-(Ymax/20), (Ymax/20)}, Axes→True, AxesOrigin→{0,0},
AxesLabel→{"x, nm", "y, nm"}, PlotRange→{-20, 20},
AspectRatio→Ymax/Xmax, ContourShading→contshad,
LabelStyle→Directive[Large, Black,
FontFamily→"Arial"],
ImageSize→{Automatic, 500}, ContourLabels→(Text[Style[
#3, 17,
FontFamily→"Arial"], {#1, #2}] &),
Contours→contset, ContourStyle→Thickness[0.002]]


ContourPlot[f[0.5, (10y)/Ymax + 0.5, (10z)/Zmax
```

```
+ 0.5][[4]]    -    ρs, {z, -(Zmax/20), (Zmax/20)}, {y,
-(Ymax/20), (Ymax/20)}, Axes→True,
AxesOrigin→{0, 0}, AxesLabel→{"z, nm", "y, nm"},
PlotRange→{-20,20},
AspectRatio→Ymax/Zmax, ContourShading→contshad,
LabelStyle→Directive[Large, Black,
FontFamily→"Arial"],
ImageSize→{Automatic, 500}, ContourLabels→(Text[Style[
#3, 17,
FontFamily→"Arial"], {#1, #2}] &), Contours→contset,
ContourStyle→Thickness[0.002]]
```

Users can easily satisfy themselves that this time period in case of the ordinary sample cell size (few mm) and thermal neutrons irradiation is of the order of nanosecond. If there is no possibility to perform long-term MDS in the scope of the one calculation within few nanoseconds it can be divided into the series of shorter ones. Hence, there is a need to merge results of the program described above working. This procedure should be equivalent averaging over all available $result$ arrays calculated for the same number of uniformly spread configurations. If the number of configurations is different from one $result$ to another you have to consider the possibility of the code modification without normalization of the $result$ (the last step in Listing 2), because this step should be made in the last turn.
As for visualization, possible way of the profiles obtained after solvent SLD subtraction visualization using s-plain functions is represented in Listing 4 and corresponding profiles presented in Fig.3.

5. Conclusion

As the reader can see from the Fig.2 there is the field of the space with the non-zero SLD value in the 1-nm vicinity of both MA and SA molecules which can influence scattering as it was mentioned in the Introduction. The consequent SANS data modeling can be performed according to the general approach or with taking into account symmetrical properties of the obtained SLD distributions (models of the scattering particles)[8]. However, this point

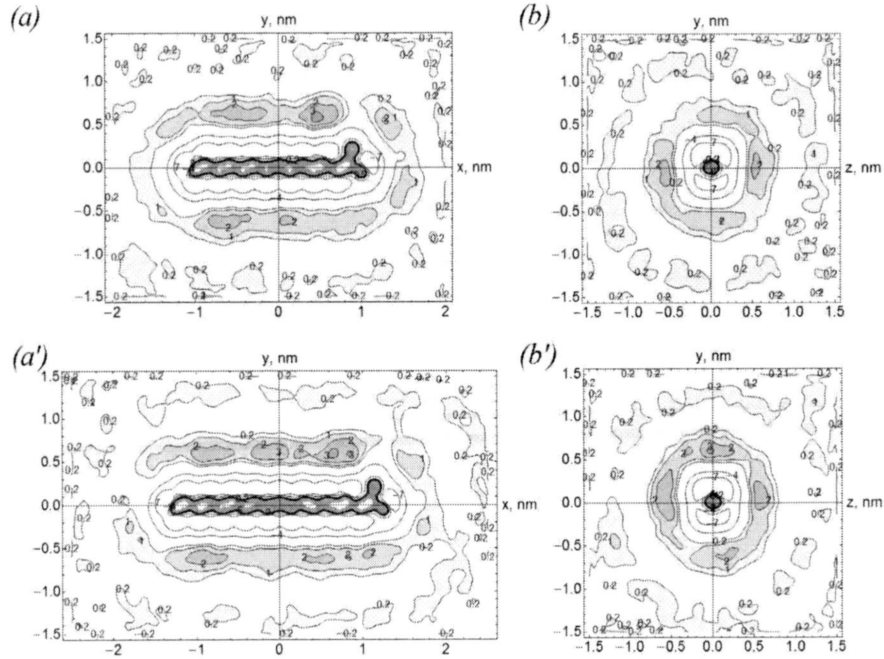

Figure 3. The longitudinal and transversal SLD spatial distributions profiles for MA (a and b, respectively) in d-decalin solution and SA(a',b') one. SLD values represented in 10^{10}cm^{-2} units.

lies beyond the scope of the current work.

Presented in the scope of the current review approach allows calculating spatial SLD distributions using trajectory files of the DL_POLY general-purpose MDS code. Authors believe that the material will be useful to specialists who have a deal with SANS experimental data resolving problem and code from Mathematica can be easily ported to any other language by readers. The work has been partially done in the frame of the RFBR projects: 12-02-00649-a, 12-02-12063-ofi_m

References

[1] Wignall, G.D., Bates, F.S., Absolute calibration of small-angle neutron scattering data. *J. Appl. Cryst.*, 20, (1987), 28-40.

[2] Petrenko, V.I., Avdeev, M.V., Almsy, L., Bulavin, L.A., Aksenov, V.L., Rosta, L., Garamus, V.M., Interaction of mono-carboxylic acids in benzene studied by small-angle neutron scattering. *J. Col. Surf.* A, 337, (2009), 91-95.

[3] Smith, W., Forester, T. R. DL_POLY_2.0: A general-purpose parallel molecular dynamics simulation package. *J. Mol. Graph.* 14, (1996), 136-141.

[4] Allen, M. P., Tildesley, D. J., *Computer Simulation of Liquids.* Clarendon Press, Oxford, (1989).

[5] Ryckaert, J. P., Bellemans, A., Molecular dynamics of liquid alkanes. Faraday Discuss. *Chem. Soc.* 66, (1978), 95-106.

[6] Eremin, R., Kholmurodov, Kh., Avdeev, M., Petrenko, V., Yasuoka, K., Molecular Dynamics Simulations on trans- and cis-Decalins: The Effect of Partial Atomic Charges and Adjustment of Real Densities. *Int. J. Chem.* 4, (2012), 14-22.

[7] Eibl, S., Fitch, A., Brunelli, M., Evans, A.D., Pattison, P., Plazanet, M, Johnson, M.R. Alba-Simionesco, C., Schober, H. trans-Decahydronaphthalene (decalin) from powder diffraction data. *Acta Cryst.* 65, (2009), 278-280.

[8] Feigin, L. A., Svergun, D. I., *Structure Analysis by Small-Angle X-Ray and Neutron Scattering.* Plenum Press, New York, (1987).

In: Models in Bioscience and Materials Research ISBN: 978-1-62808-052-0
Editor: Kh. T. Kholmurodov

Chapter 11

Computational Design of Small Peptide Inhibitors of Protein-Protein Interactions in CRK-SH2 Mediated Intracellular Signaling

Junya Yamagishi[1,2,*]***, Noriaki Okimoto***[2]***, Takuma Kasai***[3]***, Atsushi Suenaga***[4]***, Mariko Okada***[5]***, Akira Imamoto***[6]
and Makoto Taiji[1,2]
[1]Department of Computational Biology
Graduate School of Frontier Sciences
The University of Tokyo, Japan
[2]Laboratory for Computational Molecular Design
QBiC, RIKEN, Japan
[3]Systems and Structural Biology Center, RIKEN, Japan
[4]Molecular Function Team, CBRC, AIST, Japan
[5]Computational and Experimental System Biology Group
RIKEN, Japan
[6]Ben May Department for Cancer Research
University of Chicago, USA

[*]E-mail address: yamagishi@riken.jp (Corresponding Author).

Abstract

Computational design of peptide inhibitors is still difficult, because conventional *in silico* methods have been developed to design drug-like small molecules. In this study, we developed docking-based protocols for the first-round screening. We developed our new docking program based on Molecular Mechanics and an implicit solvent model. We used a GPGPU technique to accelerate our calculations. We also considered the ligand's reorganization effect. This effect corresponds to the energetic contribution of the conformational change of the ligand through binding to the target protein. We used a small training set of phosphopeptides against CRK SH2 domain, which is a target for new cancer therapy. We showed high screening performance in discriminating between binding and non-binding sequences. Our methods will also be helpful for more accurate design of small molecule inhibitors in the future.

Keywords: Docking Study, CRK SH2 Domain, GPGPU, Reorganization Effect, Peptide Inhibitor, Protein-Protein Interactions

1. Introduction

CRK (CT10 Regulator of Kinase) protein is a member of an adaptor protein family. Adaptor proteins have no enzymatic activity, but they contain several protein-binding modules. Adaptor proteins mediate specific protein-protein interactions: they drive the formation of signaling protein complexes through binding of each signaling component to CRKs binding modules. CRK have two types of protein-binding domains, Src Homology 2 (SH2) domain and SH3 domain. SH2 domains recognize phosphorylated regions of the binding partner, while SH3 domains recognize proline-rich regions. Previous studies revealed that pYXXP is a consensus binding motif for CRK SH2 domain [1]; pY represents a phosphorylated tyrosine residue and X represents any amino acids. CRK is known to be indispensable for malignant features of human cancers [2]. Thus, CRK SH2 and SH3 domains are attractive targets for new cancer therapy.

Peptides are believed to be potent agents to inhibit or modulate interactions between signaling proteins. The binding ability of peptides is guaranteed because the peptide, or a part of the protein, is an original ligand of protein-protein interactions. Actually, several binding peptide sequences binding to CRK SH2 or SH3 domains were identified experimentally [3, 4]. So the design of peptide inhibitors seems to be easier than designing small molecule inhibitors directly.

Peptide design, however, arise other problems due to peptide-specific characteristics.

In silico molecular design techniques can estimate binding affinities of unknown ligands to their target proteins. A structure-based approach is a first choice for CRK SH2 and SH3 domain, because a sequence-based approach requires a large amount of experimental information. The structure-based approach uses a 3D-structure of a target protein determined experimentally or computationally to estimate binding affinities. The structure-based approach is classified into two groups according to their computational costs and accuracies: one group uses a single structure or a few (multiple) structures. This approach places emphasis on a speed of calculation so it is used for first-round screening. Molecular docking is one of the major techniques, and is one of the main subjects of this work. While the other approach uses dynamic ensemble of 3D-structures and it can predict binding affinities more accurately. Recently, Molecular Dynamics simulation (MD) based approaches are mainly used in this area.

In this study, we applied the docking based strategy to design peptide inhibitors for CRK SH2 domain. We tried to modify conventional protocols specific for peptide design because molecular docking-based approaches have been developed for special use to design small molecules, in contrast to more accurate approaches which can be applicable for small-molecule and peptide design.

Peptide docking requires more complicated scoring functions to estimate the binding affinity correctly. To reduce computational costs, conventional scoring functions pick up only typical interactions between the ligand and the target protein like hydrogen-bonds or ion-interactions. This is good approximations for drug-like small molecules because they have a few hydrogen donors/acceptors and charged moieties. Thus, their bindings to the target protein are dominated by several typical interactions. On the other hand, peptides have many polar functional groups. Therefore it is important to take into consideration the interaction not only with the protein but with surrounding waters. An implicit solvent model used in Molecular Mechanics will be a useful approximation to take account into solvent effects, but this is unsuited for docking calculations due to the high computational demand .

To overcome these difficulties, we decided to develop our own docking program for peptide docking. We selected an all-atom force field potential energy function and generalized born (GB) implicit solvent as a scoring function [5]. To meet the high computational demands of our scoring function, we took advantage of Graphical Processing Unit (GPU) acceleration. GPUs are specialized for highly parallel computations. In the docking simulation, each binding pose evaluation is completely independent and the force field and GB calculations in each pose evaluation can be easily parallelized. Therefore the force field and GB-based docking is well suited for GPU computation.

Alternatively, flexible positional restraints need to be implemented for peptide docking. Many rotatable bonds of peptides make their conformational space very large, and make it difficult to search binding poses correctly. Therefore positional restraints based on a reference complex structure are useful to reduce their conformational spaces.

In addition to the prediction of peptide-protein complexes, predictions of the peptide conformations in solution are considered to be very important in the peptide design. We can easily expect peptides to change their conformations more significantly than small molecules through binding to the protein. Energetic contributions by these conformational changes are non-negligible, especially for short peptide ligands which has no specific stable conformations. These effects are known as ligands reorganization effects or ligands restraint energy. We performed additional conformational searches of peptides with the implicit solvent (without the protein), and incorporate these effects into the binding affinity prediction.

The purpose of this study is to incorporate new features described above into our docking-based methods and to examine our screening performance with a small training set. For the future, we are planning to design peptides de novo or based on experimentally determined sequences using a genetic algorithm. Our work serves as a basis for the future peptide design.

2. Methods

2.1. Training Set

Previous studies revealed both binding and non-binding sequences of peptides to CRK SH2 domain experimentally [3, 4]. All peptide sequences contain a phosphorylated tyrosine residue, but their length varies widely. We cut all sequences into eight amino acids: two residues from pY to the N-terminal and 5 residues from pY to the C-terminal (XXpYXXXXX). This small training set contains 38 pYXXP binding sequences, 28 pYXXP NON-binding sequences and 74 NON-pYXXP NON-binding sequences.

The screening performance was digitalized by values of the area under curve of the ROC curve (ROC AUC).

2.2. CRK SH2 Conformations

Generally, the performance of our screening based on molecular docking is sensitive to the conformation of the receptor protein. Thus we need to compare the performances using a variety of conformations to find an optimal one. We generated the conformational ensemble of the peptide-CRK SH2 complex by MD simulation. AMBER ff99SB and modified force field for the phosphorylated-tyrosine were used [6]. Initial structure (PDB: 1JU5) was solvated by TIP3P waters box. We performed the MD simulation by sander program in AMBER 8 with a RIKEN MD-GRAPE3 special purpose computer [7]. After 2 ns equilibration at 300K, 100 conformations were sampled every 200 ps.

2.3. Development of Our Docking Software

We developed our own docking program with CUDA language. We used a free energy of a whole system as a scoring function. Our docking score, DS, is expressed as:

$$DS = E_{MM} + G_{GB}$$

where E_{MM} is the AMBER ff99SB potential energy function and G_{GB} is the generalized born solvation energy. Any cutoff scheme was performed in calculations of non-bond interactions. A soft 8-6 Lennard-Jones potential function was used to permit some steric clashes. The soft Lennard-Jones potential function is expressed as:

$$E_{LJ} = \epsilon(3(\frac{r_m}{r})^8 - 4(\frac{r_m}{r})^6)$$

where ϵ is the well depth, r is the distance between two particles and r_m is the distance at which the potential reaches its minimum. Coefficients were determined to maintain the same energy at r_m.

We added some positional restraints to the ligand while docking. We used the conformations of the phosphopeptide extracted from peptide-CRK SH2 domain complex generated by MD as a reference. The side chain atoms of the phosphorylated-tyrosine were fixed at the positions of the reference. Additionally, we applied harmonic restraints on the Carbon-alpha atoms from -1(one residues from pY to the N-terminal) to +4 residues (four residues from pY to the C-terminal). If the distance between the carbon alpha atom of the docked pose and that of the reference structure is greater than 2.0 Å, the harmonic restraint energy was add to DS with the force constant $10\ \mathrm{kcal/mol} \cdot \mathrm{Å}^2$. All receptor atoms were fixed through whole process.

We used the genetic algorithm to optimize binding poses from initial conformations generated randomly. We implemented peptide-docking three times per one sequence and generated 10 best poses from each runs. A total of 30 conformations were retained for the next energy minimization procedure.

2.4. Conformational Search of the Peptide without Protein

We performed the additional conformational search of peptides without CRK SH2 domain to predict the stable conformation in solution. The GPU program described above was also used for this calculation. In the same manner as the prediction of complex structure, a total of the 30 conformations were retained for the next energy minimization step.

2.5. Structural Refinements Using the Energy Minimization Algorithm

All predicted structures were refined using the energy minimization algorithm. All structures were solvated by TIP3P waters, and all TIP3P waters were removed after the structural refinement. The convergence criterion, dRMS, was set to $0.5\ \mathrm{kcal/mol} \cdot \mathrm{Å}$ and $0.3\ \mathrm{kcal/mol} \cdot \mathrm{Å}$ for the complex and the ligand, respectively.

2.6. Single MM-PBSA Rescoring

We performed MM-PBSA rescoring of refined structures. A free energy of a system, G, is expressed as:

$$G = E_{MM} + G_{PB} + G_{SA}$$

where G_{PB} is a polar solvation free energy and G_{SA} is a nonpolar solvation free energy. E_{MM} and G_{SA} were calculated by sander program in AMBER11. G_{PB} was calculated by apbs 1.3 [8]. The details of the calculation of G_{PB} followed Swansons work [9]. PB radii were set by a BONDI radii set with offset (+0.373).

The binding affinity ΔG_{bind} was defined as:

$$\Delta G_{bind} = G_{com} - (G_{rec,c} + G_{lig,c})$$

where G_{com} is G of the most stable complex structure of the 30 conformations. $G_{rec,c}$ and $G_{lig,c}$ are defined as G of CRK SH2 domain and G of phosphopeptide extracted from the most stable complex structure, respectively. The binding affinity including the ligands reorganization effects is expressed as:

$$\Delta G_{bind} = G_{com} - (G_{rec,c} + G_{lig,w})$$

where $G_{lig,w}$ is G of the most stable phosphopeptide structure in solution.

3. Results

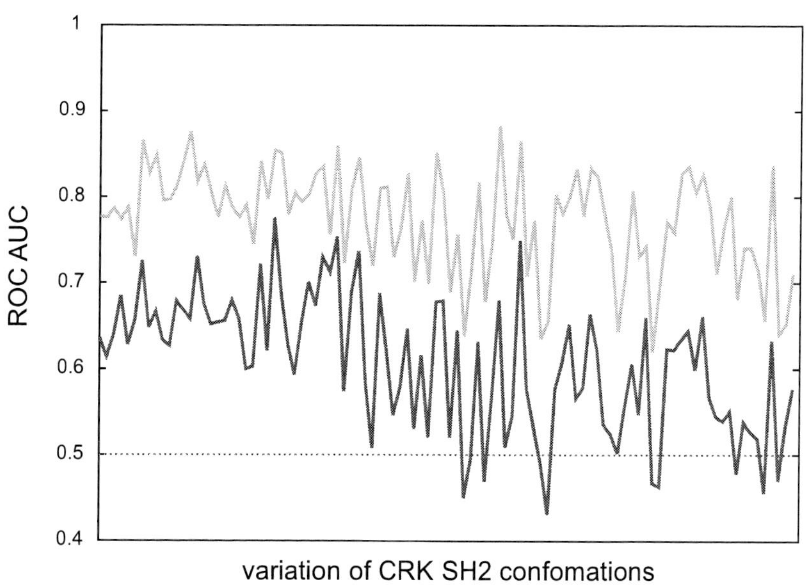

Figure 1. Values of ROC AUC associated with 100 receptor conformations. A green line includes ligand's reorganization effects, while a red one does not include that effects.

The graph in Figure 1 shows values of ROC AUC associated with each reference conformation, where the positives were defined as pYXXP binding sequences. In this graph, a green line is based on the ΔG_{bind} including ligands reorganization effects and red one based on ΔG_{bind} NOT including these effects. You will see that values of the red line always exceed that of red one. From this, ligands reorganization effects have proven to be very helpful in discriminating binding sequences from a small training set.

The graph in Figure 2 also shows values of ROC AUC. A green line also based on the ROC curve where the positives are defined as pYXXP binding sequence, but a red one based on the ROC curve where positives corresponds to pYXXP NON-binding sequences. Both of them include ligands reorganization effects. You will see that values of ROC AUC for NON-binding pYXXP sequences keep low. This means that our screening methods can detect only binding sequences from both binding and non-binding sequences having pYXXP core motif.

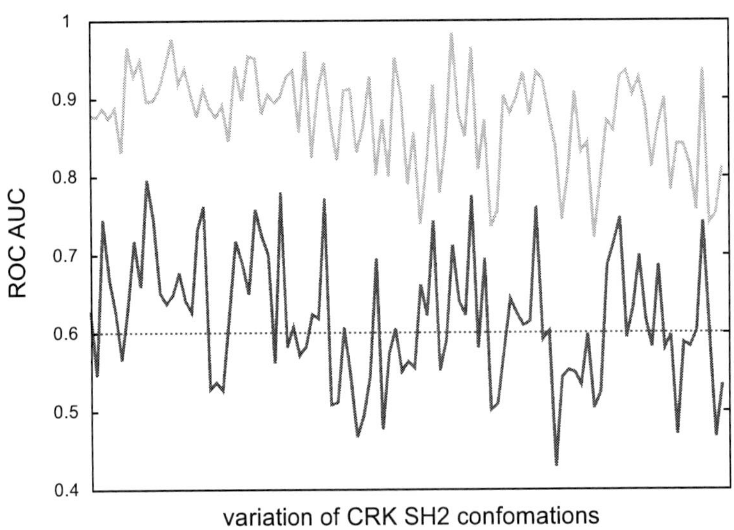

Figure 2. Values of ROC AUC associated with 100 receptor conformations. Both of two lines includes ligand's reorganization effects. A green line based on the ROC curve where positives corresponds pYXXP binding sequences, while in a red one based on the ROC curve where positives correspoinds pYXXP NON-binding sequences.

Figure 3 shows the ROC curve having the highest value of AUC 0.883, where positives were defined as pYXXP binding sequences. A red line based on the ΔG_{bind} including ligands reorganization effects, while a green one not including that effects. You can see a remarkable improvement at initial increases which is important for *in silico* screening.

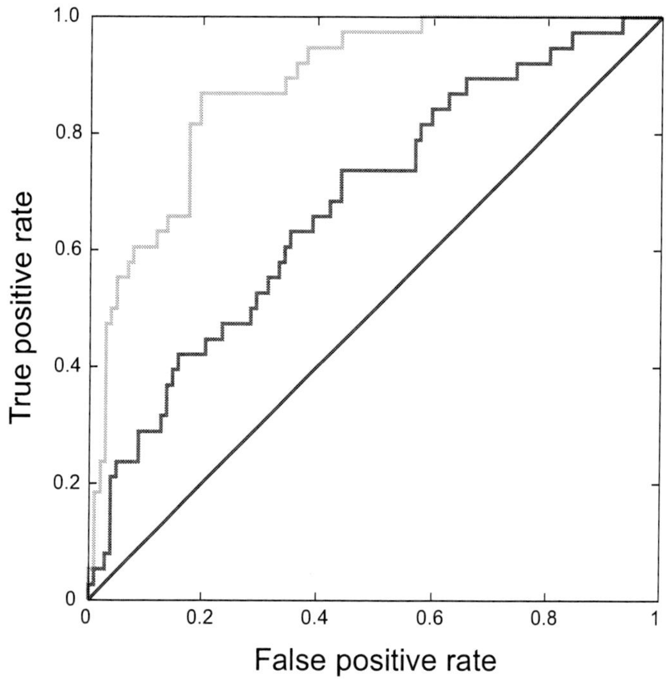

Figure 3. ROC curve having the best screening performance. A green line includes ligand's reorganization effects, while a red one does not include that effects. A blue one based on random sampling.

Acknowledgments

All calculations using GPU were carried out on the TSUBAME2.0 supercomputer in the Tokyo Institute of Technology. Other calculations were mainly performed by using the RIKEN Integrated Cluster of Clusters (RICC) facility. The development of the MD-GRAPE3 special purpose computer was supported by Research Revolution 2002 "Protein 3000 Project", operated by MEXT. This work was supported by Grant-in-Aid for JSPS Fellows 11J06408. A special thanks goes to Ms. Airi Chujo at Ritsumeikan University.

References

[1] Songyang Z., Shoelson S. E., McGlade J., et al. Specific motifs recognized by the SH2 domains of Csk, 3BP2, fps/fes, GRB-2, HCP, SHC, Syk, and Vav. *Mol. Cell. Biol.* 1994; 14: 2777-2785.

[2] Linghu H., Tsuda M., Makino Y., et al. Involvement of adaptor protein Crk in malignant feature of human ovarian cancer cell line MCAS. *Oncogene*, 2006; 25: 3547-3556.

[3] Liu B. A., Jablonowski K., Shah E. E., et al. SH2 domains recognize contextual peptide sequence information to determine selectivity. *Mol. Cell. Proteomics.* 2010; 9: 2391-2404.

[4] Watanabe T., Tsuda M., Makino Y., et al. Crk adaptor protein-induced phosphorylation of Gab1 on tyrosine 307 via Src is important for organization of focal adhesions and enhanced cell migration. *Cell Res.* 2009; 19: 638-650.

[5] Onufriev A., Bashford, and D., Case D. A. Exploring protein native states and large-scaleconformational changes with a modified generalized Born model. *Proteins*, 2004; 55: 383-394.

[6] Homeyer N., Horn A. H. C., Lanig H., et al. AMBER force-field parameters for phosphorylated amino acids in different protonation states: phosphoserine, phosphothreonine, phosphotyrosine, and phosphohistidine. *J. Mol. Model.* 2005; 12: 281-289.

[7] Taiji M. MDGRAPE-3 chip: a 165 Gflops application specific LSI for molecular dynamics simulations. *IEEE Computer Society.* 2004; in CDROM .

[8] Baker N. A., Sept D., Joseph S., et al. Electrostatics of nanosystems: application to microtubules and the ribosome. *Proc. Natl. Acad. Sci.* USA 2001; 98: 10037-10041.

[9] Swanson J. M. J., Wagoner J. A., Baker N. A., et al. Optimizing the Poisson Dielectric Boundary with Explicit Solvent Forces and Energies: Lessons Learned with Atom-Centered Dielectric Functions. *J. Chem. Theory Comput.* 2007; 3: 170 -183.

In: Models in Bioscience and Materials Research ISBN: 978-1-62808-052-0
Editor: Kh. T. Kholmurodov

Chapter 12

MOLECULAR DYNAMICS SIMULATIONS OF THE DNA INTERACTION WITH METALLIC NANOPARTICLES AND TiO_2 SURFACES

Kholmirzo T. Kholmurodov[1,2,*]***, Ermuhammad B. Dushanov***[1,3]***,***
Evgenii A. Krasavin[1,2]***, Hagar K. Hassan***[4]***, Hadeer A. ElHabashy***[5]***,***
Ahmed Galal[4]***, Nasser H. Sweilam***[6] ***and Kenji Yasuoka***[7]
[1]Laboratory of Radiation Biology, JINR, Dubna, Russia
[2]Dubna International University, Dubna, Russia
[3]Institute of Nuclear Physics, Ulugbek, Uzbekistan
[4]Chemistry Department, Faculty of Science, Cairo University, Egypt
[5]Physics Department, American University in Cairo, Egypt
[6]Mathematics Department, Faculty of Science, Cairo University, Egypt
[7]Department of Mechanical Engineering, Keio University, Japan

Abstract

The understanding of the mechanism of DNA interactions and binding with metallic nanoparticles (NPs) and surfaces represents a great interest in today medicine applications due to diagnostic and treatment of oncology diseases. Recent experimental and simulation studies involve the DNA interaction with highly localized proton beams or metallic NPs (such as Ag, Au, etc.), aimed on targeted cancer therapy through the injection of metal micro- or nanoparticles into the tumor tissue with conse-

*E-mail address: mirzo@jinr.ru (Corresponding Author)

quent local microwave or laser heating. The effects of mutational structure changes in DNA and protein structures could result in destroying of native chemical (hydrogen) bonds or, on the contrary, creating of new bonds that do not normally exist there. The cause of such changes might be the alteration of one or several nucleotides (in DNA) or the substitution of specific amino acid residues (in proteins), that can brought to the essential structural destabilization or unfolding. At the atomic or molecular level, the replacement of one nucleotide by another (in DNA double-helices) or replacement of one amino acid residue by another (in proteins) cause essential modifications of the molecular force fields of the environment that break locally important hydrogen bonds underlying the structural stability of the biological molecules. In this work, the molecular dynamics (MD) simulations were performed on four DNA models and the flexibilities of the purine and pyrimidine nucleotides during the interaction process with the metallic NPs and TiO_2 surface were clarified.

Keywords: DNA, Nanoparticles, Surfaces, Titanium Dioxide, Mutation Effects, Molecular Dynamics Simulations

1. Introduction

The effects of mutational structure changes in biological objects (like as DNA, proteins and so on) have mostly associated with different kinds of diseases that take place and develop at physiological conditions. In human body, for example, the origins of many diseases correlate with the malfunctioning of DNA or proteins, followed by their structural perturbation, improper conformational behavior or unfolding. The induced mutations in biological molecules (DNA, proteins, etc.) have quite a different nature (environmental factors and viruses, ionizing radiation and mutagenic chemicals, inherited genetic alterations, and so on). Point mutations in DNA and protein structures could result in destroying of native chemical (hydrogen) bonds or, on the contrary, creating of new bonds that do not normally exist there. The cause of such changes might be the alteration of one or several nucleotides (in DNA) or the substitution of specific amino acid residues (in proteins), that can brought to the essential structural destabilization or unfolding. At the atomic or molecular level, the replacement of one nucleotide by another (in DNA double-helices) or replacement of one amino acid residue by another (in proteins) cause essential modifications of the molecular force fields of the environment that break locally important hydrogen bonds underlying the structural stability of the biological molecules. As a result, we get global structure changes of the biomolecule, which make their functional behaviour different than that of the native ones. In other words, a specific disease (say, cancer) can develop if the biomolecule (DNA, protein) become unable to perform their function [1-3].

High-power beams of particle accelerators (say, like as JINR nuclear facilities in Dubna, Russia) provide a wide basis for the investigation of the mutagenic effect of ionizing radiation. The radiation genetics and radiobiology research objects include the mechanism of the induction of mutations of different nature by ionizing radiation. In some aspects, computer molecular simulation and analysis appear to be an extremely efficient tool for supporting various biophysical, biochemical or radiobiological experiments. Specially mentioned should be the modern molecular dynamics approach, which is widely applied in today's biophysics, radiobiological and materials science research. In the present work, based on the advanced molecular dynamics methods and visualization techniques we simulate the behaviour of native and mutated DNA interacting with metallic nanoparticles and surface [1-13].

Using molecular dynamics (MD) simulation technique we were aimed to elucidate the important DNA–metal interaction mechanism in two closely correlated aspects:

1. The DNA–metallic nanoparticles (NPs) binding;

2. The DNA–metallic oxide surface interaction.

Firstly, the MD study one performs to investigate the physical and molecular mechanism of the interaction of nano- and microparticles with DNA (a native double-helix chain and its several mutated versions – on the one end, on minor and major grooves). It is worth noting that understanding the mechanism of DNA–NP (nanoparticle) interaction represents a great interest in today medicine applications due to diagnostic and treatment of oncology diseases. Some recent experimental and simulation studies involve the interaction of DNA with highly localized high power beams and various nanoparticles (Ag, Au, etc.), which were aimed on targeted cancer therapy through the injection of metal micro- or nanoparticles into the tumor tissue with consequent local microwave or laser heating. One has to mention some activities on using of nuclear medicine facilities operating at JINR that are performed for the above purposes by the JINR's Laboratory of Radiation Biology (under the guidance of Academy corr.-mem. Krasavin E.A.).

Secondly, one extend the DNA–nanoparticles interaction to DNA–surface one by simulating the interaction of DNA with metal oxide (titanium dioxide, TiO_2). It is well-known that titanium dioxide (TiO_2) has a valuable variety of applications in a lot of different areas such as heterogeneous catalysis and photocatalysis, sensoristics, cosmetics, sunscreen, toothpaste, coating, etc. Ti as a metal has a good biocompatibility due to unique characteristics such as lightness, excellent mechanical properties, lack of toxicity, and extremely low corrosion rate. These properties make it suitable for the use as a medical implant and prosthesis material [14]. Titanium surface reacts spontaneously with air and biological solutions forming an inert thin oxide layer (0.5-10 nm thick) that is crucial for molecular recognition [15]. Titanium dioxide (TiO_2) is a naturally occurring metal oxide mineral while TiO_2 nanoparticles are manufactured artificially in large quantities. However, the synthesis and applications of titanium dioxide nanoparticles (TIO_2 NPs) are increasing due to their distinctive physico-chemical characteristics, and increased industrial and medical applications. The consistently expanding usage of TiO_2 and TiO_2 NPs has evoked urgent concern about their potential health effect. Although, TiO_2 is chemically inert, TiO_2 NPs can cause negative health effects so that the biological responses to NPs may exceed those of micron-sized particles. Because of their small size which do not exceed a hundred nanometers at maximum, high number per given mass, large specific surface area, and generation of free radicals, NPs can penetrate cells and influence several subcellular mechanisms. Some studies report that nanoparticles may defuse and penetrate into cell nuclei and may directly or indirectly affect the structure and function of DNA [16, 17, 18, 19].

TiO_2 NPs may directly or indirectly damage DNA through interactive oxidative stress and/or inflammatory responses. One *in vivo* study illustrate that TiO_2 NPs induces genotoxicity, oxidative DNA damage, and inflammation in a mice model while other *in vitro* showed that it can produce a significant induce cyto- and genotoxicity in cultured Wistar Institute, Susan Hayflick (WISH) cells. On one hand, some studies proved that there is a kind of direct chemical interaction between TiO_2 NPs and DNA, through the DNA phosphate group, but no direct evidence for the mutagenesis. On the other hand, TiO_2 NPs nanoparticles can cause indirect DNA damage through inflammation and generation of reactive oxygen species (ROS) that mainly the main reason for tumor induction. Additionally, in [20] it was revealed that, TiO_2 NPs decreases the reduced nicotineamide adenine dinucleotide (NADH) levels and impairs membrane potential and mitochondrial function accompanied by ROS generation during mi-

Table 1. The force field potentials and parameters for TiO_2 surface

Buckingham potential: $A_{ij}\exp(-r_{ij}/\rho_{ij}) - C_{ij}/r_{ij}^6$			
$l-j$	A_{ij} (kJ/mol)	ρ (Å)	C_{ij} (kJ/mol·Å^6)
Ti–Ti	3004646.13432	0.154	506.54433
Ti–O	1637244.37188	0.194	1214.7449
O–O	1137621.80484	0.234	2915.7794

tochondrial respiration. TiO_2 has recently been reclassified by the international agency for research on cancer (IARC) as group 2B carcinogen. In [21] it is mentioned that genotoxicity studies that measure different types of DNA damage (e.g., gene mutations, chromosomal damage, and DNA strand break formation) are an important part of cancer research and risk assessment of potential carcinogens. Thus, mutation in the recognition part of a DNA generally causes alterations in its protein binding affinity that alter on their physiological functions. The mutation may cause considerable structural change particularly the groove width in the DNA as well as the interactions at the DNA-protein interface may be weakened considerably due to mutation [16, 17, 18, 19, 22, 23, 24, 25].

2. Materials and Methods

A classical molecular dynamics study was performed using the DL_POLY_4.03.3 [26] general-purpose code. The NVT ensemble at $T = 298$ K with a Nosé-Hoover thermostat and Verlet leapfrog scheme were employed. To calculate the long-ranged electrostatic forces, three dimensional Ewald summation with the automatic optimisation parameter $f = 1.0 \cdot 10^{-4}$ and convergence parameter 0.24375 Å^{-1} was used. The integration time step of the dynamic equations of motion was 1 fs.

All the initial and mutated DNA configurations were created using the Amber Tools LEaP basic preparation program [4, 5]. The DNA was solvated with water (TIP3P) molecules [27] produced in spherical and rectangular (periodic) water baths. In creating the force field parameters the combination of CHARMM and "general Amber force field (GAFF)" databases [4, 5, 28] were employed.

The DNA–water system were equilibrated at normal (room) temperatures and located between two TiO_2 (anatase) surfaces. The MD cell volume of the

DNA–water–TiO_2 system was as $V = (37.85, 56.775, 57.639)$ Å^3. (The only DNA–water volume was 70463.018 Å^3; its total mass 28431.872 a.m.u.; so the density was equal to 0.670 g/cc).

The water atoms were constrained using the SHAKE method [29].

For the TiO_2 surfaces, the force fields as reported by Kavathekar et al. [30] and Guillot et al. [31] were used. The potential parameters for TiO_2 surface were developed by Matsui and Akaogi [32]. Unit cell parameters for TiO_2 model were taken from the EIM databases supported by the Russian Foundation for Basic Research [33]. Table 1 represents the Buckingham (buck) potential parameters used for the TiO_2 surface [32].

The results of simulations and images of the molecular systems were analyzed by the Visual Molecular Dynamics (VMD) [7] software.

3. Results and Discussion

The MD simulation results below cover the following model systems: (1) The DNA–metallic nanoparticles (NPs) binding; (2) The DNA–metallic oxide surface interaction. The details of the MD simulations are given in "Materials and Methods" section.

3.1. The DNA–Metallic NP interactions

First, we have generated a spherical water bath embedding the DNA and a small metallic NP. The DNA and water (TIP3P) model parameters were taken from the standard AMBER database. As for the NP the force field parameters for the gold atoms were used. We have generated several DNA–water–NP configurations with single, two, three–five atomic sites to mimic various small nanoparticle clusters. In Fig. 1 a snapshot of the DNA chain and metallic nanopartcile (NP) surrounded by water is presented.

We have generated multiple DNA–water–NP configurations and equilibrated all models at a normal (room) temperature of $T = 300$ K. The initial positions of the DNA and NP inside the water bath were well-separated and well-distinguished. The total molecular systems under considerations next to undergo a long-nanosecond relaxation procedure up to 10-12 ns (nanoseconds). In Fig. 2 six sequential snapshots on the DNA molecule interacting with NP are presented.

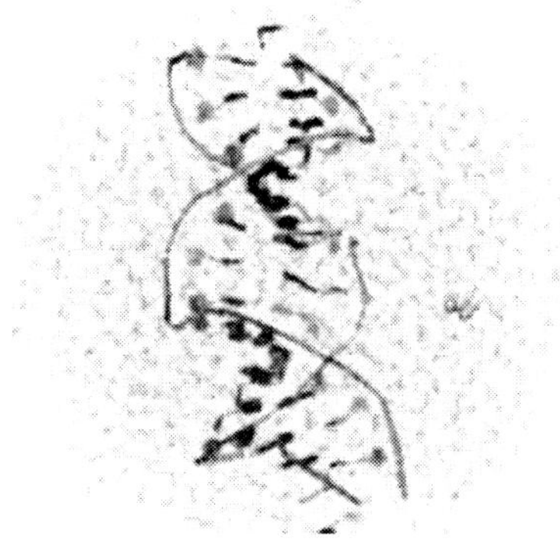

Figure 1. A DNA chain embedded by water and interacting with metallic nanopartcile (NP). The DNA shown as ribbon structure, water – as a blue bath; NP – as golden balls.

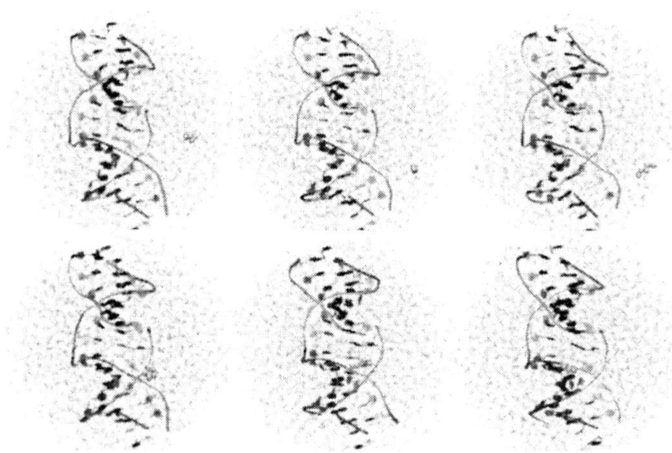

Figure 2. Six sequential snapshots on DNA–water–NP interaction process (from left to right). The DNA conformation behaves stable till the NP is capture by a DNA minor groove.

During the long-nanosecond dynamics the metallic NPs would continuously change their positions, approaching the DNA molecule from its different sides (from the DNA ends, the DNA minor and major grooves, etc.), thereby seeking a preferrable binding site with. One of the events of the DNA–NP binding has graphically illustrated in Fig. 3.

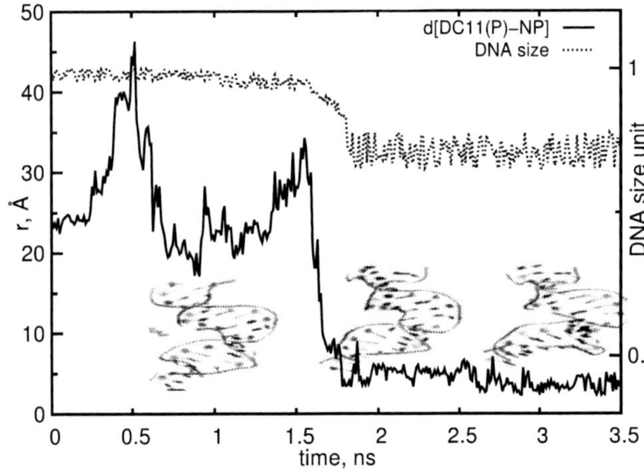

Figure 3. A distance diagram displaying the NP trajectories and formation of DNA–NP close contact bonding. The DNA shown as ribbon structure, NP – as golden balls. Right Y-axis demonstrates the DNA longitudinal size deformation correlated with the capture of NP by a DNA minor groove.

Until NP binds to DNA, the DNA conformation behaves stable and it doesn't changes visibly in comparison with the DNA starting one. In Fig. 3 we can see that within of $t \approx 1.5 - 2$ ns from the start of the equilibration dynamics the NP is localized at a DNA minor groove. From this stage of a DNA–NP close contact (bond) formation the DNA structure undergoes a substantial deformation. The shape of the DNA molecule would continuously changes in correlation with the NP capturing and formation of a close contact with. In Fig. 3 the changes on a relative DNA unit size are shown separately (right Y-axis); it is seen that the dynamical changes on the DNA size following by the capture of NP vary from 17% up to 26%.

The process of the DNA dynamical size change is presented separately on the Fig. 4, where the DNA conformational modifications after the capture of NP at DNA minor groove are illustrated.

Based on the above observations, one conclude that different parts of the DNA molecule have likely to possess a different flexibility with regard to an efficient capturing of metallic NPs and interaction with. Say, from the above results one observe that an efficient NP–DNA interaction has preferably localized on a DNA minor groove segment. So, next we peformed the MD simulations

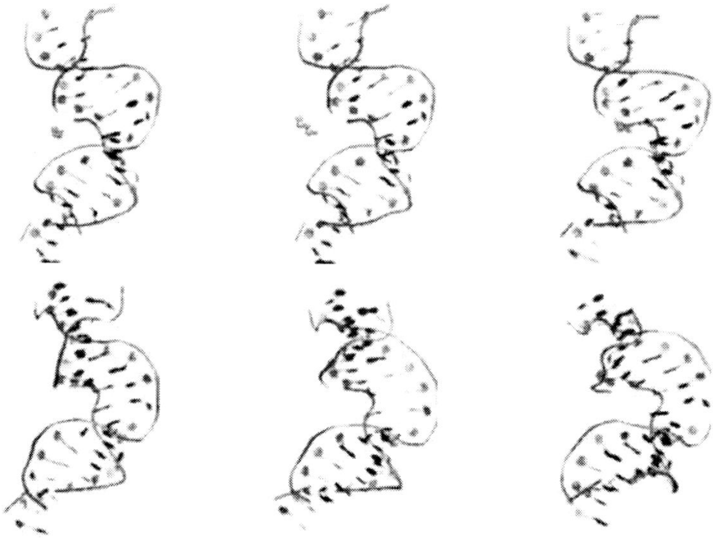

Figure 4. Six sequential snapshots of the DNA interacting with NP (from left to right). The process of the DNA conformation changes with NP capture is illustrated.

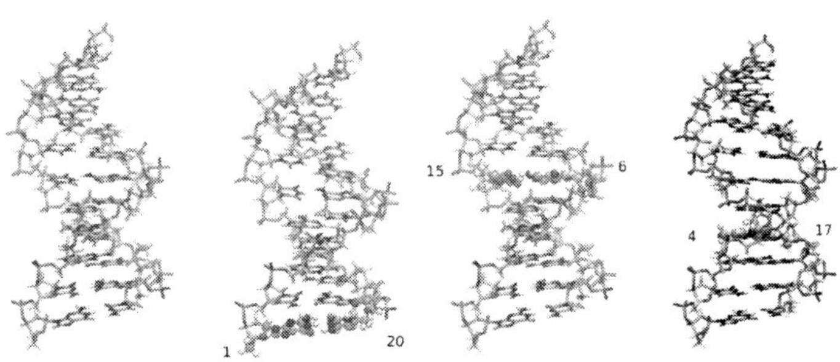

Figure 5. Four models of DNA simulated as in Figs. 1-2 with water. From the left to right: model `m0` – native structure; model `m1` – G1C and C20G (G/C→C/G) mutations; model `m2` – A6C and T15G (A/T→C/G) mutations; model `m3` – T4G and A17C (T/A→G/C) mutations.

and some enegetics estimations on the only DNA chain solvated in water, but without any nanoparticle (NP) inclusion into consideration. Fig. 5 and Table 2

Table 2. The DNA models described as in Fig. 5 are numbered and sequenced as following: Model m0: no mutation, Model m1: G1C and C20G mutations (G/C→C/G), Model m2: A6C and T15G mutations (A/T→C/G), Model m3: T4G and A17C (T/A→G/C) mutations

DNA chain numbering:	1	2	3	4	5	6	7	8	9	10
	11	12	13	14	15	16	17	18	19	20

Model m0:	Model m1:	Model m2:	Model m3:
5'-GCGTTAACGC-3'	5'-**C**CGTTAACGC-3'	5'-GCGTT**C**ACGC-3'	5'-GCG**G**TAACGC-3'
3'-CGCAATTGCG-5'	3'-**G**GCAATTGCG-5'	3'-CGCAA**G**TGCG-5'	3'-CGC**C**ATTGCG-5'

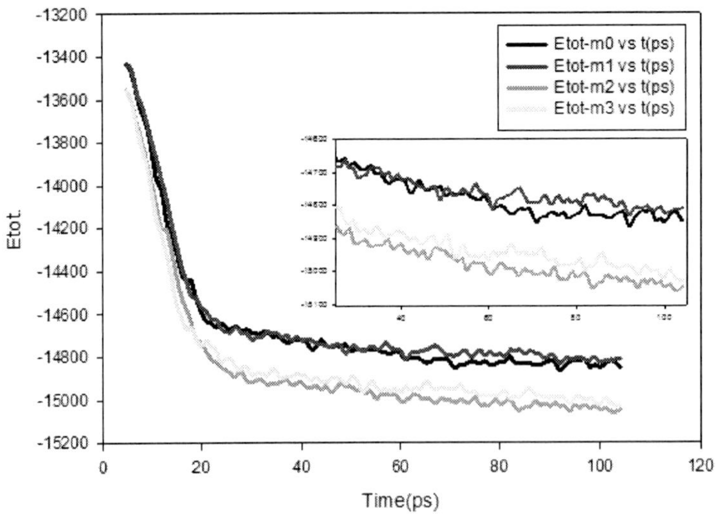

Figure 6. This graph shows the total evaluated energies from the MD simulations results for four different models of solvated DNA system versus.

show four DNA models simulated with water as in Figs. 1-2. We have introduced these DNA models as: model m0 – native structure; model m1 – G1C and C20G mutations; model m2 – A6C and T15G mutations; model m3 – T4G and A17C mutations, so far to compare the flexibility of different segments of the DNA molecule (the DNA ends, minor and major grooves). For better understanding of the influence of the induced mutations with respect to the DNA native one, we have run molecular dynamics simulations on the four DNA models, each solvated with the same number of water molecules. In Fig. 6 the total energy vs time is drawn for each of the DNA–water models. The two curves in Fig. 6 demonstrate that models `m0` (native DNA) and `m1` (DNA chain mutated on the one end) have a comparable energetics behavior. At the same time, the third and the forth DNA-water models, `m2` and `m3`, show a relative significant decrease in energy than the native `m0` and `m1` ones. One have to stress out that mutations in the models `m2` and `m3` are located at the DNA major and minor grooves, respectively.

Let us estimate the thermodynamic stabilities of these four DNA models with respect to the induced mutations as shown in Table 2. It's well known that thermodynamic stability of DNA double helix depends on its length where longer molecules are more stable, its GC-content and sequence since stacking is sequence specific. The four DNA models under investigation are ten bases pairs in length, so their energetic stabilities due to length dependency are expected to be similar and cannot be used a point of comparison [19]. On the other hand hydrogen bonds between base pairing nucleotides are the main reason where higher GC-content (%) is energetically favorable for DNA molecules. That's because different base pairing nucleotides form different numbers of hydrogen bonds. AT base pair forms two hydrogen bonds while GC forms three hydrogen bonds. DNA with high GC-content is more thermodynamic stable and would have lower energy than DNA with lower GC-content. Calculating the GC-content percentages of our models, we have found that the first native DNA, `m0`, and the chain end mutated DNA, `m1`, have the same GC-content which is 60%,while the third major groove mutated DNA, `m2`, and the forth minor groove mutated DNA, `m3`, have the same GC-content which is 70%. Both the GC base pairs content and the overall length of a DNA can determine the strength of the association between the two strands of DNA. Since the four DNA models have the same length, models with higher GC-content have stronger-interacting strands among other models. Moreover calculating predicted values of the total energies of our models individually considering the nuclobases pairing and stacking (attractive, non-covalent interactions between aromatic rings) upon one another to form the helical structure of DNA will be done using the nearest neighbor (NN) parameters for Watson-Crick base pairs in 1 M NaCl. The published Watson-Crick NN parameters have been critically reviewed many times in literature and have great reliability. The total Gibbs free energy for the models can be calculated using the following equation

$$\Delta G^0_{37(tot)} = \Delta G^0_{37(int)} + \Delta G^0_{37(sym)} + \sum \Delta G^0_{37(st)}, \tag{1}$$

where $\sum \Delta G^0_{37(st)}$ is the summation of the standard free-energy changes for the 10 possible Watson-Crick NNs and $\Delta G^0_{37(int)}$ accounts for differences between duplexes with terminal AT vs. terminal GC pairs, $\Delta G^0_{37(sym)}$ equals 10.43 kcal/mol if the duplex is self-complementary and equals zero if it is non-self-complementary. All estimated values of the energy have been measured at 37 Celsius degree the physiological temperature of human which. It's clear that the values of $\Delta G^0_{37(int)}$ and $\Delta G^0_{37(sym)}$ for our non-complementary structures will be the same for the four models and $\sum \Delta G^0_{37(st)}$ will be the main reason of the energy fluctuation between the four models [16, 18].

The calculated values of the $\sum \Delta G^0_{37(st)}$ for the four DNA models are summarized below.

The first native model `m0` (no mutation):

```
5'-GCGTTAACGC-3'
3'-CGCAATTGCG-5'
```

$$\begin{aligned}
\Delta G^0_{37(tot,m0)} &= \Delta G^0_{37(int)} + \Delta G^0_{37}\binom{G\,C}{C\,G} + \Delta G^0_{37}\binom{C\,G}{G\,C} + \Delta G^0_{37}\binom{G\,T}{C\,A} + \Delta G^0_{37}\binom{T\,T}{A\,A} + \\
&+ \Delta G^0_{37}\binom{T\,A}{A\,T} + \Delta G^0_{37}\binom{A\,A}{T\,T} + \Delta G^0_{37}\binom{A\,C}{T\,G} + \Delta G^0_{37}\binom{C\,G}{G\,C} + \Delta G^0_{37}\binom{G\,C}{C\,G} + \\
&+ \Delta G^0_{37(sym)} = 1.96 - 2.24 - 2.17 - 1.44 - 1.00 - 0.58 - 1.00 - 1.44 - 2.17 - \\
&- 2.24 + 0.00 = -12.32 \text{ kcal/mol}
\end{aligned}$$

The chain end mutated model `m1` (mutated G1C and C2G):

```
5'-CCGTTAACGC-3'
3'-GGCAATTGCG-5'
```

$$\begin{aligned}
\Delta G^0_{37(tot,m1)} &= \Delta G^0_{37(int)} + \Delta G^0_{37}\binom{C\,C}{G\,G} + \Delta G^0_{37}\binom{C\,G}{G\,C} + \Delta G^0_{37}\binom{G\,T}{C\,A} + \Delta G^0_{37}\binom{T\,T}{A\,A} + \\
&+ \Delta G^0_{37}\binom{T\,A}{A\,T} + \Delta G^0_{37}\binom{A\,A}{T\,T} + \Delta G^0_{37}\binom{A\,C}{T\,G} + \Delta G^0_{37}\binom{C\,G}{G\,C} + \Delta G^0_{37}\binom{G\,C}{C\,G} + \\
&+ \Delta G^0_{37(sym)} = 1.96 - 1.84 - 2.17 - 1.44 - 1.00 - 0.58 - 1.00 - 1.44 - 2.17 - \\
&- 2.24 + 0.00 = -11.92 \text{ kcal/mol}
\end{aligned}$$

The major groove mutated model `m2` (mutated A6C and T15G):

```
5'-GCGTTCACGC-3'
3'-CGCAAGTGCG-5'
```

$$\begin{aligned}\Delta G^0_{37(tot,m2)} &= \Delta G^0_{37(int)} + \Delta G^0_{37}\binom{G\,C}{C\,G} + \Delta G^0_{37}\binom{C\,G}{G\,C} + \Delta G^0_{37}\binom{G\,T}{C\,A} + \Delta G^0_{37}\binom{T\,T}{A\,A} + \\ &+ \Delta G^0_{37}\binom{T\,C}{A\,G} + \Delta G^0_{37}\binom{C\,A}{G\,T} + \Delta G^0_{37}\binom{A\,C}{T\,G} + \Delta G^0_{37}\binom{C\,G}{G\,C} + \Delta G^0_{37}\binom{G\,C}{C\,G} + \\ &+ \Delta G^0_{37(sym)} = 1.96 - 2.24 - 2.17 - 1.44 - 1.00 - 1.28 - 1.45 - 1.44 - 2.17 - \\ &- 2.24 + 0.00 = -13.47 \text{ kcal/mol}\end{aligned}$$

The minor groove mutated model m3 (mutated T4G and A17C):

5'-GCG**G**TAACGC-3'
3'-CGC**C**ATTGCG-5'

$$\begin{aligned}\Delta G^0_{37(tot,m3)} &= \Delta G^0_{37(int)} + \Delta G^0_{37}\binom{G\,C}{C\,G} + \Delta G^0_{37}\binom{C\,G}{G\,C} + \Delta G^0_{37}\binom{G\,G}{C\,C} + \Delta G^0_{37}\binom{G\,T}{C\,A} + \\ &+ \Delta G^0_{37}\binom{T\,A}{A\,T} + \Delta G^0_{37}\binom{A\,A}{T\,T} + \Delta G^0_{37}\binom{A\,C}{T\,G} + \Delta G^0_{37}\binom{C\,G}{G\,C} + \Delta G^0_{37}\binom{G\,C}{C\,G} + \\ &+ \Delta G^0_{37(sym)} = 1.96 - 2.24 - 2.17 - 1.84 - 1.44 - 0.58 - 1.00 - 1.44 - 2.17 - \\ &- 2.24 + 0.00 = -13.16 \text{ kcal/mol}\end{aligned}$$

The predicted energy values with the nearest neighbor (NN) parameters for Watson-Crick base pairs in 1 M NaCl gave us a general view about the energetic and stability statuses of the models. Fortunately, the above MD simulation results for the four DNA molecules give us matching outputs with the predicted orders for the four models energies. One have to note out that there is an approximate constant shift in energy values evaluated from MD simulations than that evaluated from the calculations. One relate it to the water path energy contribution in the total energy of the DNA-water system. To prove this assumption we have run molecular dynamics for the water molecules alone with the same particles number and simulation parameters as used for solvated DNA models. That matches positively with our assumption that the water energy has approximate constant shift effect on the total energy of the four solvated DNA models. Thus, the predicted energy values and the evaluated energy from the MD simulation results are in agreement that model m2 has the lowest energy, $\Delta G^0_{37(tot,m2)}$ and the highest thermodynamic stability among the other models followed by model m3 $\Delta G^0_{37(tot,m3)}$ then model m0, $\Delta G^0_{37(tot,m0)}$, finally model m1 $\Delta G^0_{37(tot,m1)}$ has the highest energy and the lowest stability among the other models. The native DNA (model `m0`) and the DNA chain end mutated (model `m1`) structures have close predicted energy values. The DNA `m2` and `m3` models have relatively lower energy values than that of `m0` and `m1` ones. So far, the DNA major and minor groove mutated models seem to possess the most favorable energetics characteristics for the specific stacked sequence under consideration. Such energetics behavior one beleive that make a sense due to DNA – metallic NPs binding phenomena as reported above. The DNA mutated sites may cause an essential modification of the DNA – NPs interaction; obviously, this might be important to be taken into account for a number of similar reactions in the applied or experimental realizations.

3.2. The DNA–Metallic Oxide Surface Interactions

In this section, we extend the DNA – metallic NPs interaction to DNA – surface one by simulating the interaction of DNA with metal oxide (titanium oxide, TiO_2).

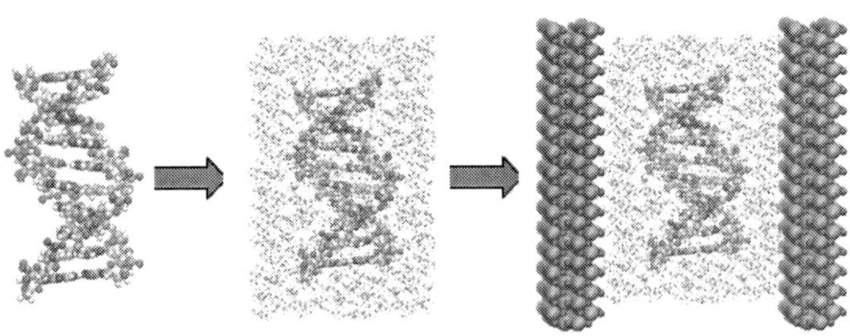

Figure 7. The initial DNA configuration (left) embedded in water (middle) and surrounded (right) in between two symmetrically TiO_2 surfaces (walls) are shown. The titanium (Ti) atoms shown as large green balls; oxygen (O) atoms – as small red balls.

Four DNA structures as described in previous section were solvated by water and located in between two symmetrically TiO_2 surfaces as shown in Fig. 7. The DNA–water–TiO_2 models were equilibrated at a room temperature under similar environmental and thermodynamical parameters (see "Materials and Methods") and after reaching the equilibrated states we have picked up for each of four models the DNA–surface distance distribution statistics (Figs. 8-11).

Model m0: no mutation

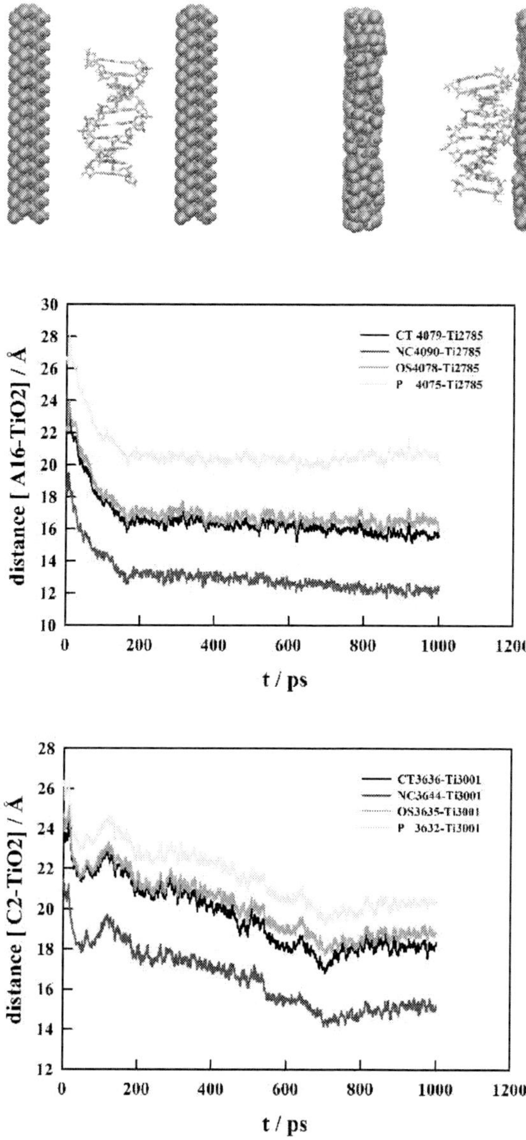

Figure 8. Continued on next page.

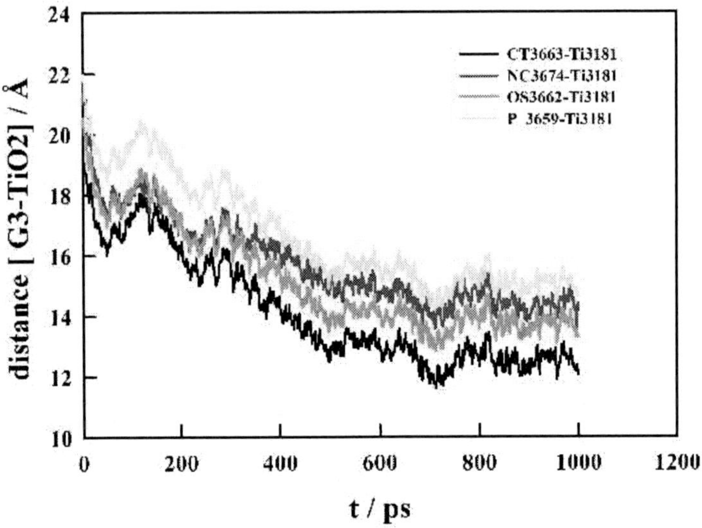

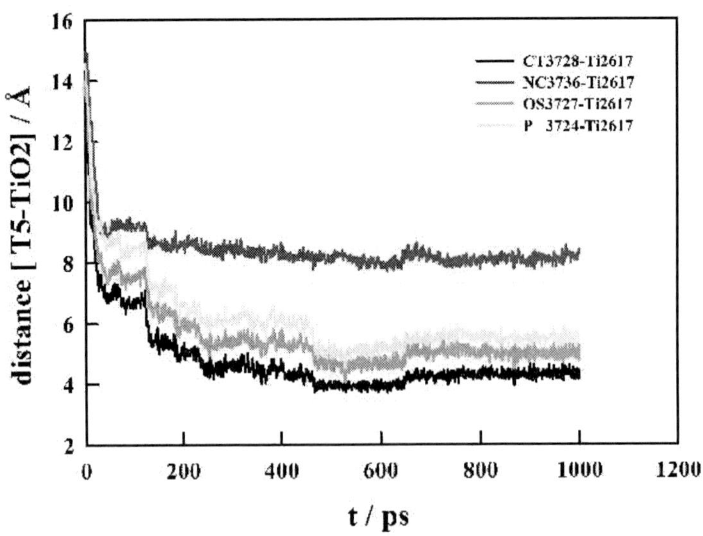

Figure 8. Continued on next page.

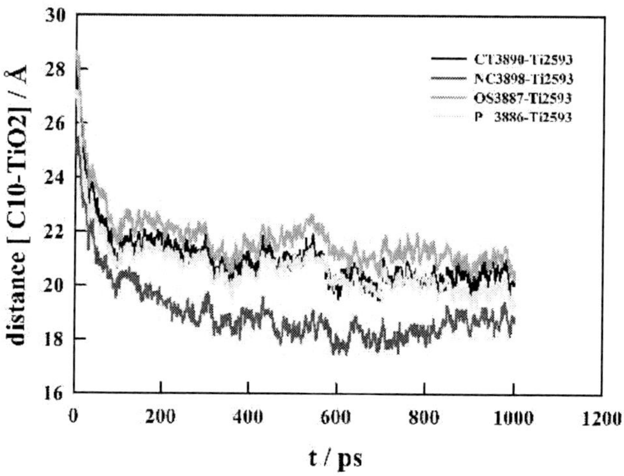

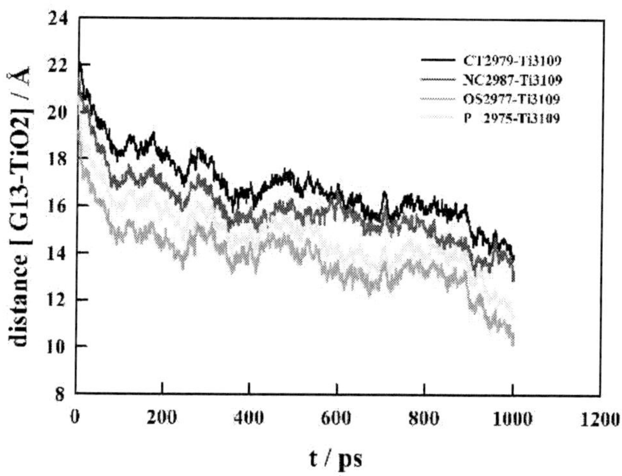

Figure 8. The initial and final DNA configurations for Model m0 (no mutation) as well as the RMSD graphs.

In Figs. 8-11 the DNA–TiO_2 distance graphs were calculated by fixing a certain Ti atom on TiO_2 surface and measuring the distance between it and the closest backbone atoms of DNA (C, N, O and P) for each of the four nucleotides A, C, G and T. We also calculated the DNA–surface distance distributions by

Model m1: G1C and C20G mutations

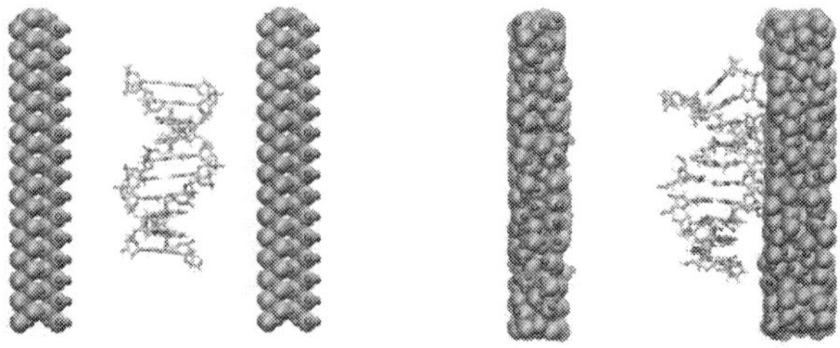

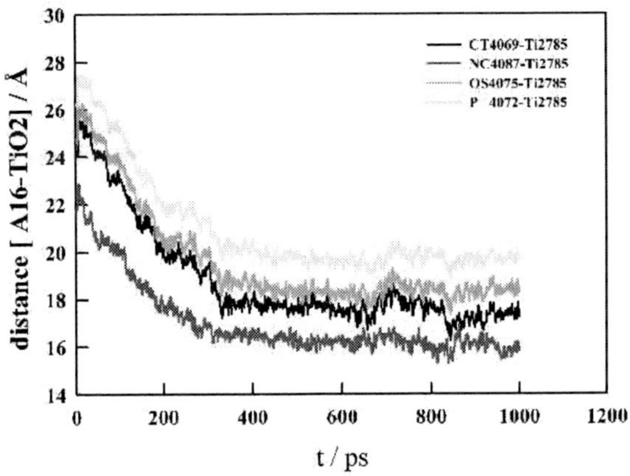

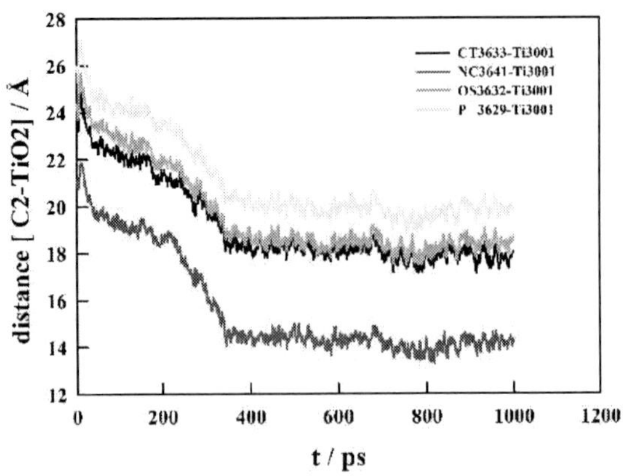

Figure 9. Continued on next page.

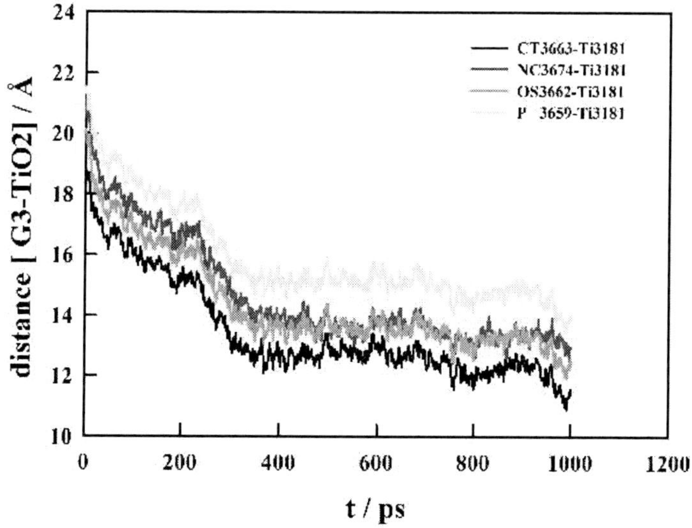

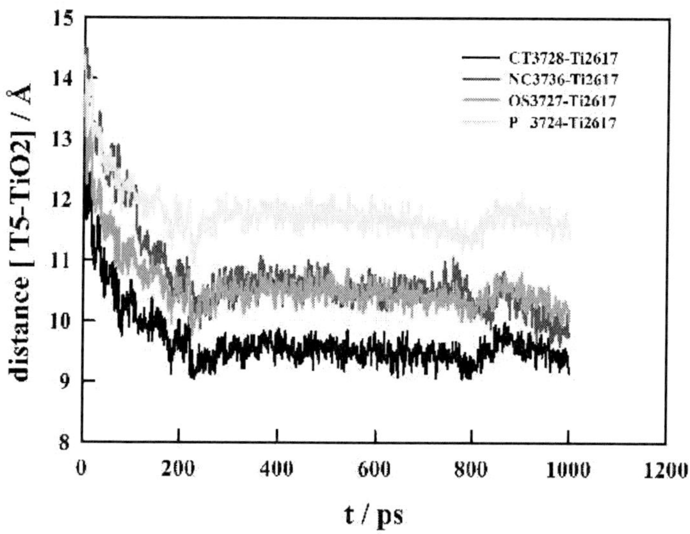

Figure 9. Continued on next page.

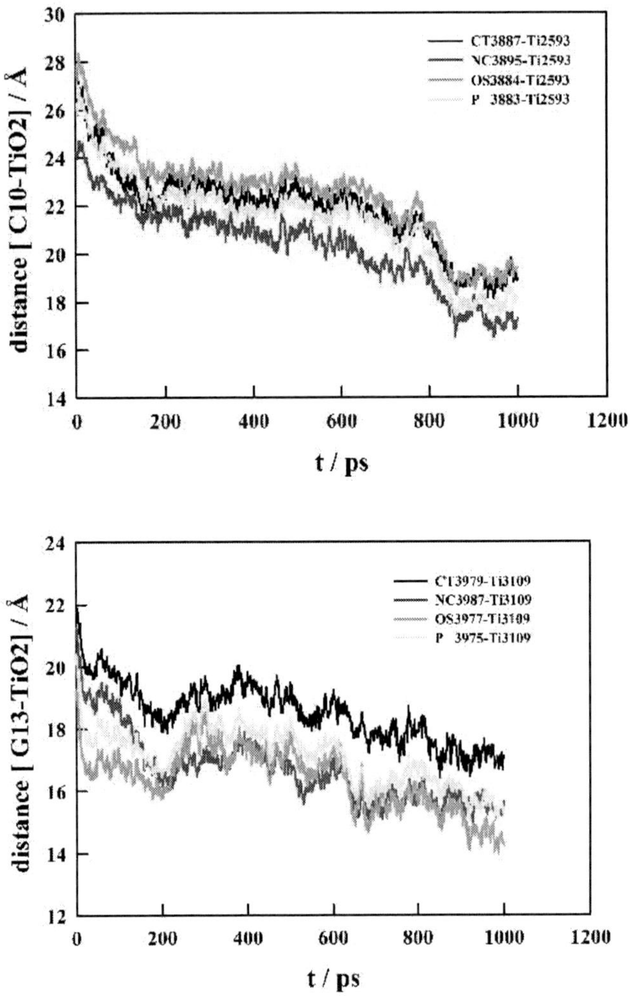

Figure 9. The initial and final DNA configurations for Model m1 (G1C and C20G mutations) as well as the RMSD graphs.

taking into account the position of the nucleotides on the DNA double strand. So that, we used three different positions for each nucleotide to predict a general picture with a lower error that may results from the uncertainties (say, due to possible fluctuations in the position changing or in nucleotide sites relative to

Model m2: A6C and T15G mutations

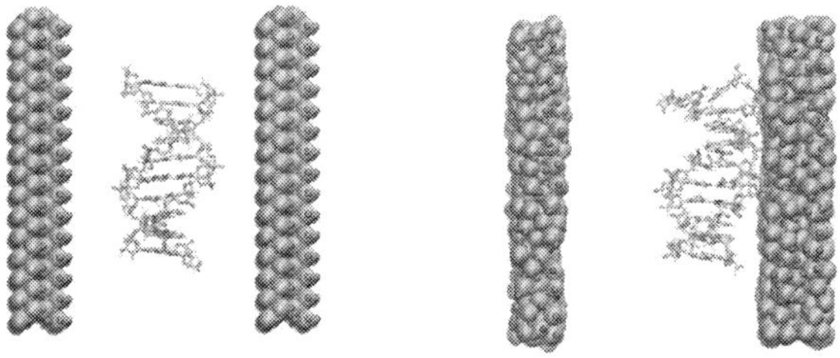

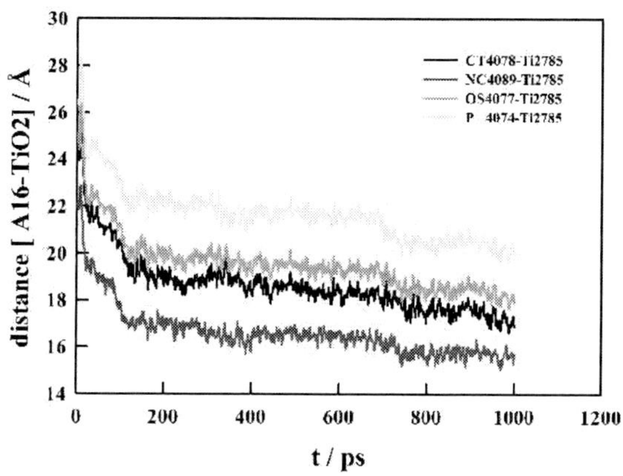

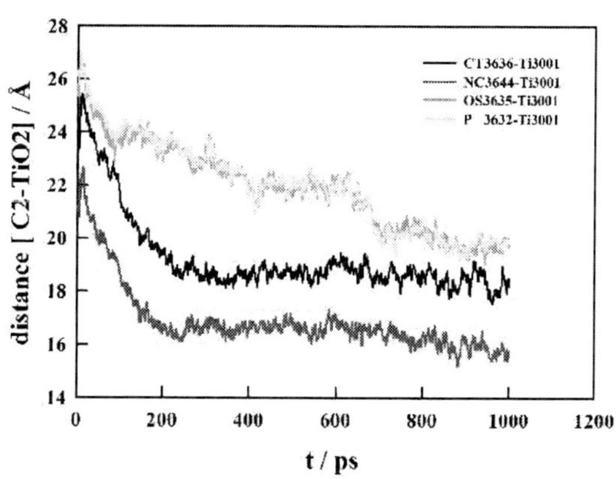

Figure 10. Continued on next page.

2D Graph 38

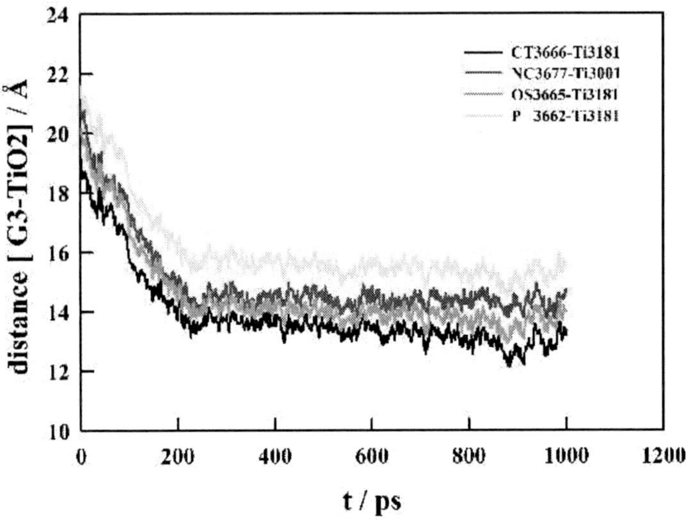

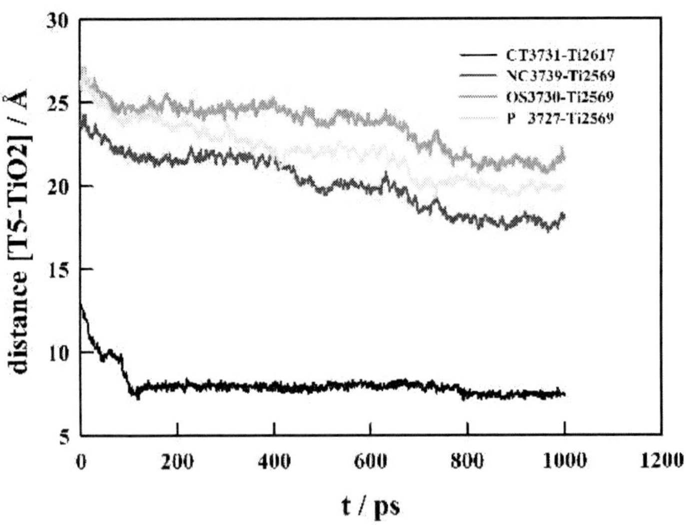

Figure 10. Continued on next page.

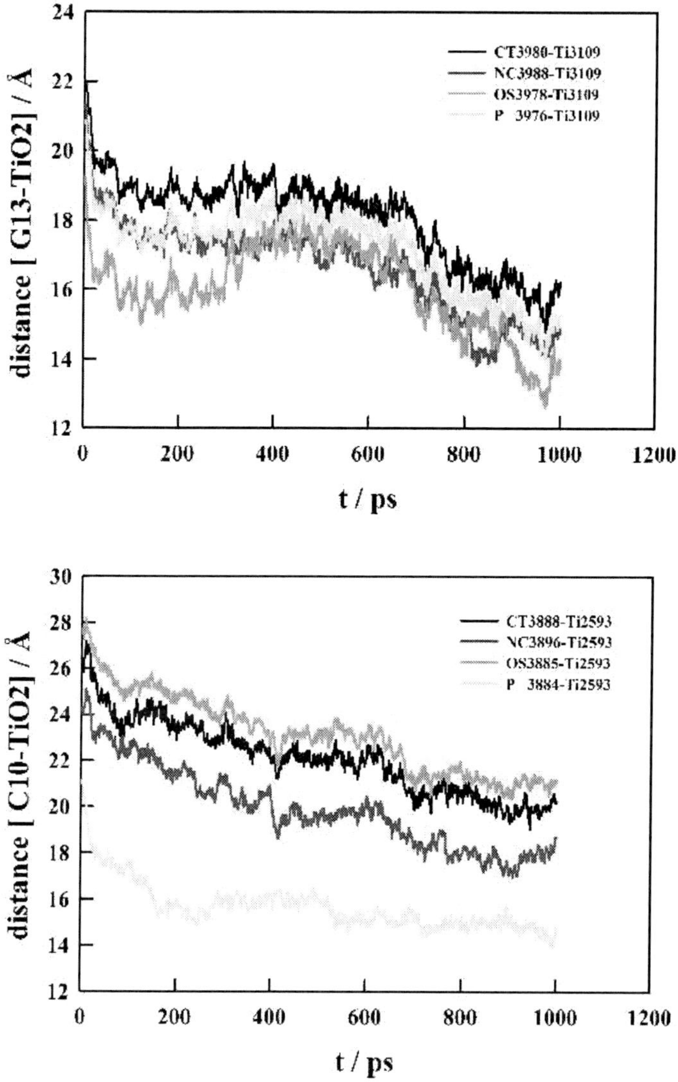

Figure 10. The initial and final DNA configurations for Model m2 (A6C and T15G mutations) as well as the RMSD graphs.

Model m3: T4G and A17C mutations

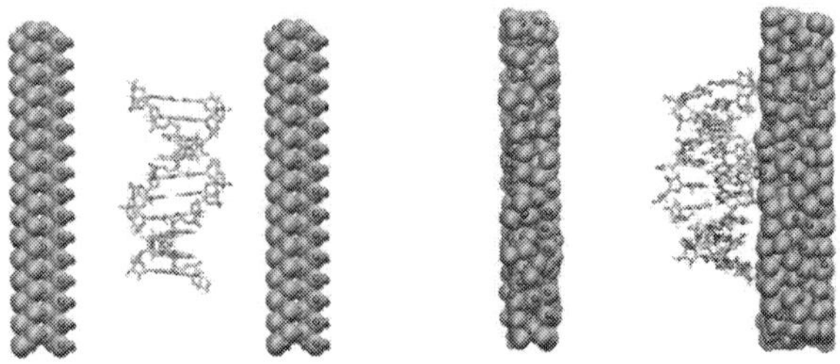

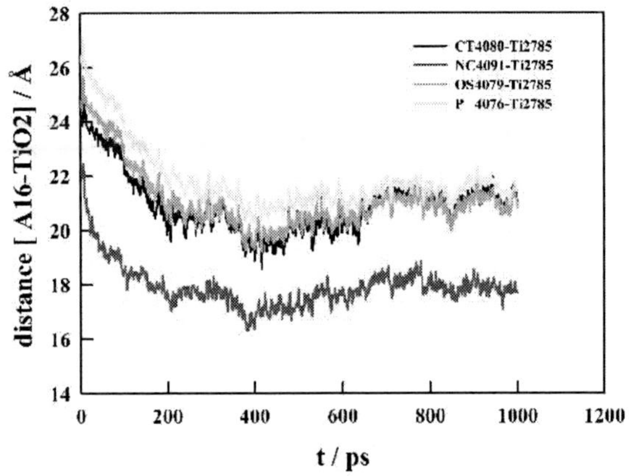

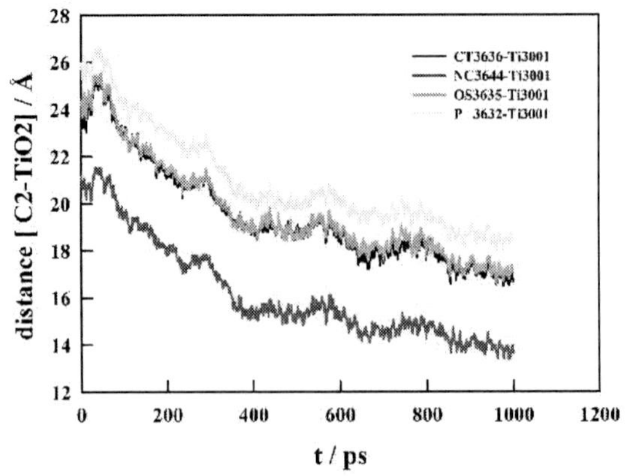

Figure 11. Continued on next page.

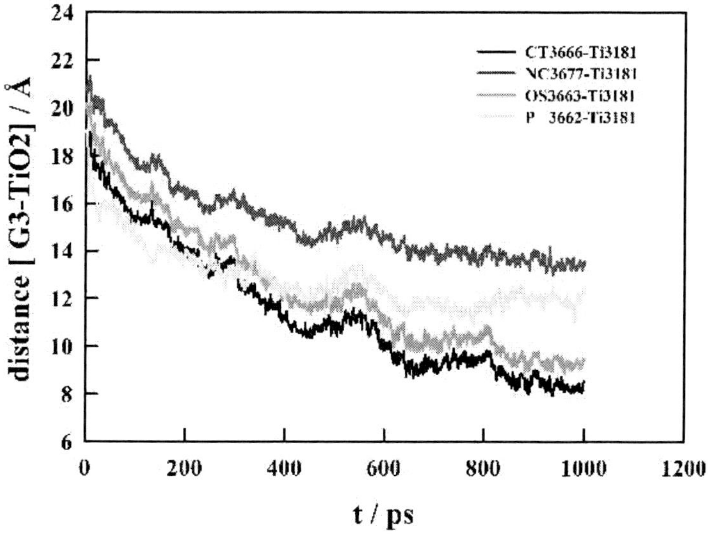

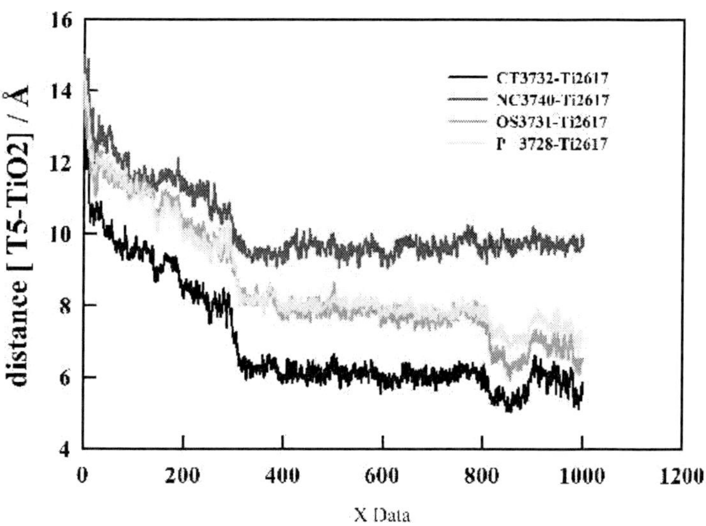

Figure 11. Continued on next page.

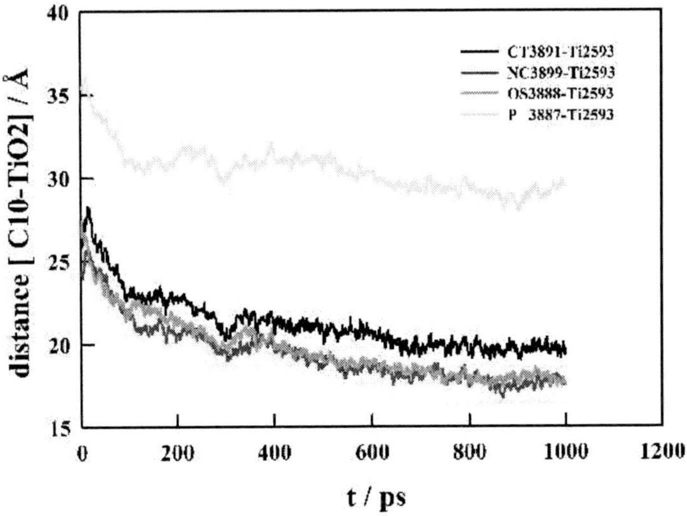

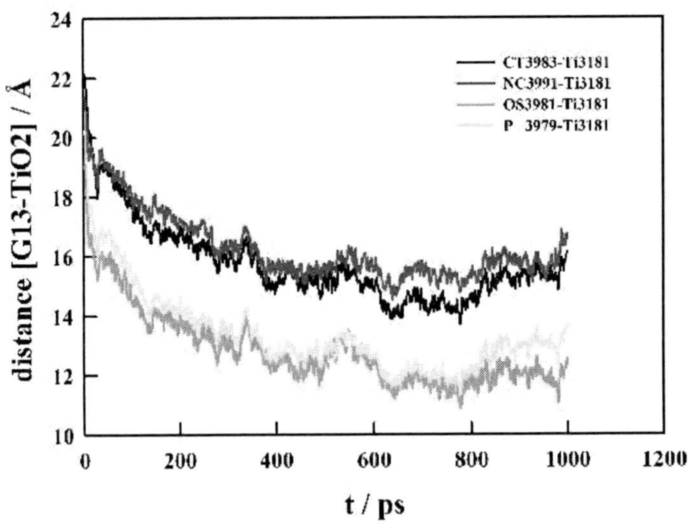

Figure 11. The initial and final DNA configurations for Model m3 (T4G and A17C mutations) as well as the RMSD graphs.

the whole DNA double strand).

In Figs. 8-11 the DNA–TiO_2 distance distributions together with the snapshots of initial and final states are presented for all simulated structures (`m0`, `m1`, `m2` and `m3`), respectively. From Fig. 8 representing the DNA–TiO_2 distances for the native structure (model `m0`) we can say that in average the closest atom to the surface is nitrogen atom which is matching with the chemistry of coordination bonding. As it is known, Ti ion in TiO_2 still has vacant orbitals which facilitate the interaction with donor atoms such as nitrogen, oxygen and phosphorous due to their lone pair of electrons. This Lewis acid- Lewis base interaction between donor atoms of DNA and Ti ion promotes some conformational changes of the DNA as predicted from the distance graphs shown in Fig. 8. The large decreasing in the distance distribution with time indicates that the structure has lost its original conformation due to the interaction with TiO_2. On the other hand, a high degree of fluctuations in the distance distribution graphs is another indicator for the conformational change of `m0` structure.

It's likely that the induced mutations on DNA structure have to affect the DNA binding ability with a TiO_2 surface. So that, the changes in the DNA–surface distances could be a good indicator to distinguish the effect of mutation. The DNA–TiO_2 distance distribution graphs for `m1`, `m2` and `m3` models are shown in Figs. 9-11, respectively. By comparing the distance distribution graphs between the three mutated structures and that of the native one we can observe some correlation on the effect of mutation (i.e. DNA conformational change due to mutation). Comparisons of Fig. 9 (model `m1`) and Fig. 8 (model `m0`) reveals that there is no big difference between these two models. The distance distribution behavior are similar for most of the nucleotides, although there seen a little fluctuation in T5, G19, etc. The largest structure changes (i.e., the highest degree of the distance distribution values) could be seen comparing the models `m0` and `m2`. From Fig. 10 (model `m2`) and Fig. 8 (model `m0`) the differences in the values of DNA–surface distances may reach 15 or 16 Å. Additionally, the `m2` closest atom to TiO_2 surface is oxygen instead of nitrogen as for `m0`. Moreover, the distance distributions have a relatively wide profile if to make comparisons between similar nucleotides. Thus, for the model `m2` one observe a relatively high structure deformation degree which is likely due to more intense DNA interaction with TiO_2 surface. From Figs. 8 (model `m0`) and 11 (model `m3`) we predict some similarities between `m2` and `m3` conformations in accordance to the mutation type in both. Estimating the DNA–TiO_2 surface interaction, based on the distance graphs Figs. 8-11, one can now summarize which structure has the highest tendency to interact with TiO_2.

Table 3. The DNA–TiO_2 distance distributions averaged between the four DNA models as described in Fig. 5 and Table 2.

Nucleotide/structure	Model m0	Model m1	Model m2	Model m3
A6	8.00	11.0	—	8.00
A7	—	—	12.0	4.00
A16	14.0	11.0	12.5	19.0
A17	14.0	14.0	12.5	—
C2	11.0	13.0	11.0	12.0
C10	10.0	11.0	14.0	17.0
C18	10.5	12.0	14.5	9.00
G3	10.0	10.0	8.00	12.5
G13	11.5	8.00	8.00	10.0
G19	9.00	11.5	9.50	8.50
T4	13.0	8.00	13.0	—
T5	19.5	5.00	19.5	9.50
T14	11.0	11.0	11.0	10.0
T15	—	—	—	13.0
Average value	11.0	10.5	12.1	11.0

From the net change in distances between the DNA and TiO_2 atoms we calculated ΔD, the difference between the upper and lower values of the close contact distanes in four DNA structures. The values of ΔD are summarized in Table 2 for all DNA models. The structure that possesses the highest value of ΔD has to be the most flexible one – the most deformable structure due to the interaction with the surface. From Table 2 the DNA structure in the mutation model m2 has the highest value of ΔD implying that DNA molecule in model m2 has probably more strongly interacts with TiO_2 surface than in the other one.

4. Conclusion

In the present paper, long-nanosecond MD simulations were performed on four DNA small chains (models m0 – m3) to elucidate the effect of induced mutations on the DNA–water–NPs and DNA–water–TiO_2 surface interactions. First, for the DNA–metallic NP interaction we have observed and energetically estimated the NP preferrable binding site to DNA structure located on its minor and major grooves. Until NP binds to DNA, the DNA conformation behaves stable and it doesn't changes visibly in comparison with the DNA starting one.

From the stage of the DNA–NP bond formation the DNA structure undergoes a substantial deformation. The shape of the DNA molecule would continuously changes in correlation with the NP capturing and formation of a close contact with.

Next we extended the DNA–NPs interaction mechanism to the DNA–TiO_2 surfaces one. We have generated four DNA structures interacting in water solvent with NP. The changes in the DNA–surface distance distributions were estimated as an indicator for the degree of DNA conformation behavior during the approaching TiO_2 surface. The major observations include the following: (a) The closest atoms to TiO_2 are Nitrogen and Oxygen as predicted due to formation of coordination bonds with Ti ion; (b) The DNA structure to be distinguished by its essential confomation changes compared with the native one has to be model `m2`; the others follow this order `m2` > `m3` > `m1`; (c) As predicted G1 $\rightarrow$ C1 mutation does not visibly affect the DNA–surface distance distributions; (d) All this mutation nearly does not affect the closest atom or in some cases nitrogen and oxygen atom alternate with each other; (e) For `m2` model, the largest increase in the DNA–surface distance profiles were observed in T5 site; the difference between the contact distances compared to that of `m0` one could reach 15 or 16 Å.

In summary, the DNA conformation has to change essentially as the nanoparticle has captured by a DNA specific site. As the DNA–nanoparticle(s) cluster together the nanoparticle(s) may pull the strands of DNA apart. The specific DNA sites that we mutated belong to a binding site that can efficiently capture the metallic nanoparticle(s). The DNA–TiO_2 interaction are likely subjected to a similar mechanism as for the DNA–nanoparticle(s) ones. So far, the DNA conformational changes under TiO_2 interactions possess a different nature with regard to four DNA modelled structures.

Acknowledgment

This work has been performed in the framework of joint collaborative agreement Arab Republic of Egypt (ARE)–Joint Institute for Nuclear Research (JINR) (project #302). This work is a part of our collaboration between JINR (Russia), RIKEN (Japan), and Keio University (Japan). The work has been performed within a research collaboration between JINR (Russia) and Keio University (Japan) and was supported by the JSPS (Japan Society for the Promotion of Science)–RFBR (Russian Foundation for Basic Research) (grant No.13-

04-92100). The work has been supported in part by the Grant in Aid for the Global Center of Excellence Program to the Center for Education and Research of Symbiotic, Safe and Secure System Design from Japan's Ministry of Education, Culture, Sport, and Technology. The MD simulations have been performed using computer software, hardware facilities, and cluster machines at the CICC (JINR), RICC (RIKEN), and the Yasuoka Laboratory of Keio University (Japan). The authors would like to specially thank Mr. Sergei Negovelov (JINR) for technical assistance and helpful comments.

References

[1] Vinall R.L, Tepper C.G., Shi X.-B., Xue L.A., Gan-dour-Edwards R. and de Vere White R.W., "The R273H p53 mutation can facilitate the androgen-independent growth of LNCaP by a mechanism that involves H2 re-laxin and its cognate receptor LGR7". *Oncogene*, 25, 2006, 2082-2093.

[2] Kholmirzo Kholmurodov (Ed.), *"Molecular Dynamics of Nanobistructures"*, Nova Science Publishers Ltd., 2011, 230p., ISBN: 978-1-61324-320-6.

[3] Kholmurodov Kh.T., Feldman T.B., and Ostrovskii M.A. Molecular Dynamics of Rhodopsin and Free Opsin: Computer Simulation. *Neuroscience and Behavioral Physiology*. 2007. 37(2), 161-174.

[4] Pearlman, D.A., Case, D.A., Caldwell, J.W., Ross, W.R., Cheatham, T.E., DeBolt, S., Ferguson, D., Seibel, G., Kollman, P., "AMBER, a computer program for applying molecular mechanics, normal mode analysis, molecular dynamics and free energy calculations to elucidate the structures and energies of molecules". *Comp. Phys. Commun.*, 91, 1995, 1-41.

[5] Case D.C., Pearlman D.A., Caldwell J.W., Cheatham III T.E., Ross W.S., Simmerling C.L., Darden T.A., Merz K.M., Stanton R.V., Cheng A.L., Vincent J.J., Crowley M., Ferguson D.M., Radmer R.J., Seibel G.L., Singh U.C., Weiner P.K., Kollman P.A., *AMBER*, 2010.

[6] Cornell, W.D., Cieplak, P., Bayly, C.I., Gould, I.R., Merz, Jr.K.M., Ferguson, D.M., Spellmeyer, D.C., Fox, T., Caldwell, J.W., Kollman, P.A., "A second Generation forth field for the simulation of Proteins and Nucleic Ac-ids". *J. Am. Chem. Soc.*, 117, 1995, 5179-5197.

[7] Humphrey, W., Dalke, A. and Schulten, K., "VMD - Visual Molecular Dynamics". *J. Molec. Graphics*, 14, 1996, 33-38.

[8] K. Kholmurodov, W. Smith, K. Yasuoka, T. Darden and T. Ebisuzaki, "A smooth-particle mesh Ewald method for DL_POLY molecular dynamics simulation package on the Fujitsu VPP700". *J. Comput. Chem.*, 21, 2000, 1187, 3.

[9] Kholmurodov Kh.T., Dushanov E.B., Yasuoka K., "MD Simulations of the P53 oncoprotein structure: the effect of the Arg273His mutation on the DNA binding domain", *Advances in Bioscience and Biotechnology*, Volume 2, Number 5, ISSN Print: 2156-8456 ISSN Online: 2156-8502 (2011).

[10] Selwyne R.A., Kholmurodov Kh.T., Koltovaya N.A., "Homology Modelling and Molecular Dynamics of Cyclin-Dependent Protein Kinases", In: *IT for Real Rorld Problems*, (Universities Press Series in Systems, Models, Informatics and Control), Ed. Sree Hari Rao (Ed.), pp. 1-72.

[11] Kholmurodov Kh.T., Krasavin E.A., Krylov V.A., Dushanov E.B., Korenkov V.V., Yasuoka K., Narumi T., Ohno Y., Taiji M., Ebisuzaki T., "MD Simulations on the Structure of Onco-Proteins P53: Wild-Type and Radioresistant Mutant Systems", In: *MOLECULAR DYNAMICS OF NANOBIOSTRUCTURES*, Nova Science Publishers (N.Y.), ISBN: 978-1-61324-320-6, 2011.

[12] Narumi T., Susukita R., Ebisuzaki T., McNiven G. and Elmergreen B., "Molecular Dynamics Machine: Special-purpose Computer for Molecular Dynamics Simulations". Molecular Simulation, 21, 1999, 401-408.; Narumi T., Susukita R., Furusawa H., Yasuoka K., Kawai A., Koishi T., Ebisuzaki T., MDM version of AMBER, 2000.; Narumi, T., Susukita, R., Furusawa, H., Ebisuzaki, T., "46 Tflops Special-purpose Computer for Molecular Dynamics Simulations: (WINE-2)". *Proc. 5th Int. Conf. on Signal Processing*. Beijing., 2000, 575-582; Susukita, R., Ebisuzaki, T., Elmegreen, B.G., Furusawa, H., Kato, K., Kawai, A., Kobayashi, Y., Koishi, T., McNiven, G.D., Narumi, T., Yasuoka, K., "Hardware accelerator for mo-lecular dynamics: MDGRAPE-2", *Computer Physics Communications*, 155, 115-131(2003).

[13] M. Taiji, T. Narumi, Y. Ohno, N. Futatsugi, A. Suenaga, N. Takada, and A. Konagaya, "Protein explorer: A petaflops special-purpose computer system for molecular dynamics simulations", in *Proceedings of SC2003*, Phoenix, AZ 2003; T. Narumi, Y. Ohno, N. Futatsugi, N. Okimoto, A.Suenaga, R.Yanai, and M. Taiji, "A high-speed special-purpose computer for molecular dynamics simulations: MDGRAPE-3", in Proceedings of NIC Workshop 2006, From Computational Biophysics to Systems Biology, *Julich*, NIC Series, vol. 34, pp.29- 35, 2006; T. Narumi, Y. Ohno, N. Okimoto, T. Koishi, A. Suenaga, N. Futatsugi, R. Yanai, R. Himeno, S. Fujikawa, M. Ikei, and M. Taiji, "A 185 TFLOPS simulation of amyloid-forming peptides from Yeast prion Sup35 with the special-purpose computer system MDGRAPE-3", in *Proceedings of SC2006*, Tampa, FL, 2006.

[14] Diebold U., "The surface science of titanium dioxide". *Surf. Sci. Rep.* 2003, 48, 53-229.

[15] Monti S., Carravetta V., Zhang W., and Yang J., "Effects Due to Interadsorbate Interactions on the Dipeptide/TiO2 Surface Binding Mechanism Investigated by Molecular Dynamics Simulations", *J. Phys. Chem.* C 2007, 111, 7765-7771.

[16] SantaLucia J.(Jr.), "A unified view of polymer, dumbbell, and oligonucleotide DNA nearest-neighbor thermodynamics", *Biochemistry*, Vol. 95, February 1998, 14601465.

[17] Tulpan D., Andronescu M., Leger S., *"Free energy estimation of short DNA duplex hybridizations, BMC Bioinformatics"*, 2010, 11:105

[18] SantaLucia J. (Jr.) and Hicks D., "THE THERMODYNAMICS OF DNA STRUCTURALMOTIFS", *Annu. Rev. Biophys. Biomol. Struct*, 2004, 415440.

[19] Breslauert K.J., Franks R., Blockers H., and Markyt L.A., "Predicting DNA duplex stability from the base sequence", *Biochemistry*, Vol. 83, June 1986, 3746-3750.

[20] Freyre-Fonseca V., Delgado-Buenrostro N.L., Gutirrez-Cirlos E.B., Caldern-Torres C.M., Cabellos-Avelar T., Snchez-Prez Y., Pinzn E., Torres I., Molina-Jijn E., Zazueta C., Pedraza-Chaverri J., Garca-Cullar C.M.,

Chirino Y.I., "Titanium dioxide nanoparticles impair lung mitochondrial function", *Toxicology Letters* 202 (2011) 111119.

[21] Trouiller B., Reliene R., Westbrook A., Solaimani P., and Schiestl R.H., "Titanium Dioxide Nanoparticles Induce DNA Damage and Genetic Instability In vivo in Mice", *Cancer Res* 2009; 69: (22). November 15, 2009.

[22] Saquib Q., Al-Khedhairy A.A., Siddiqui M.A., Abou-Tarboush F.M., Azam A., Musarrat J., "Titanium dioxide nanoparticles induced cytotoxicity, oxidative stress and DNA damage in human amnion epithelial (WISH) cells", *Toxicology in Vitro*, Vol. 26, 2012, 351361

[23] Trouiller B., Reliene R., Westbrook A., et al., "Titanium Dioxide Nanoparticles Induce DNA Damage and Genetic Instability In vivo in Mice", *Cancer Res.*,vol. 69, 2009, 8784-8789.

[24] Dunford R., Salinaro A., Cai L., Serpone N., Horikoshi S., Hidaka H., Knowland J., "Chemical oxidation and DNA damage catalysed by inorganic sunscreen ingredients", *FEBS Letters*, vol. 418, 1997, 87-90.

[25] Roy S., Sen S., "Exploring the potential of complex formation between a mutant DNA and the wild type protein counterpart: A MM and MD simulation approach", *Journal of Molecular Graphics and Modelling* 25 (2006) 158168.

[26] Todorov, I.T.; Smith, W.; Trachenko, K.; Dove, M.T. "DL_POLY_3: new dimensions in molecular dynamics simulations via massive parallelism". *J. Mater. Chem.*, 2006, 16(20), 1911–1918.

[27] Jorgensen W.L., Chandrasekhar J., Madura J.D. "Comparison of simple potential functions for simulating liquid water". *J. Chem. Phys.*, 1983, 79, 926–935.

[28] B.R. Brooks, C.L. Brooks III, A.D. Mackerell(Jr.), L. Nilsson, R.J. Petrella, B. Roux, Y. Won, G. Archontis, C. Bartels, S. Boresch, A. Caflisch, L. Caves, Q. Cui, A.R. Dinner, M. Feig, S. Fischer, J. Gao, M. Hodoscek, W. Im, K. Kuczera, T. Lazaridis, J. Ma, V. Ovchinnikov, E. Paci, R.W. Pastor, C.B. Post, J.Z. Pu, M. Schaefer, B. Tidor, R.M. Venable, H.L. Woodcock, X. Wu, W. Yang, D.M. York, and M. Karplus. "Charmm: The biomolecular simulation program". *Journal of Computational Chemistry*, 2009, 30(10), 1545–1614.

[29] Ryckaert J.P., Ciccotti G., Berendsen H.J.C. "Numerical integration of the Cartesian equations of proteins and nucleic acids". *J. Comput. Phys.*, 1997, 23, 327–341.

[30] Kavathekar, R.S.; Dev, P.; English, N.J.; MacElroy, J.M.D. "Molecular dynamics study of water in contact with the TiO_2 rutile-110, 100, 101, 001 and anatase-101, 001 surface". *Molecular Physics*, 2011, 109(13), 1649–1656.

[31] Guillot, B.; Sator, N. "A computer simulation study of natural silicate melts. Part I: Low pressure properties". *Geochimica et Cosmochimica Acta*, 2007, 71(5), 1249–1265.

[32] Matsui, M.; Akaogi, M. "Molecular dynamics simulation of the structural and physical properties of the four polymorphs of TiO_2". *Molecular Simulation*, 1991, 6(4-6), 239–244.

[33] WWW-MINCRYST. *"Crystallographic and crystallochemical database for minerals and their structural analogues"*. 2012. Available from `http://database.iem.ac.ru/mincryst/`

INDEX

E

F

G

H

I

J

K

L

M

N

O

P

Q

R

S